REGIONAL AND URBAN GIS

Regional and Urban GIS

A Decision Support Approach

Timothy L. Nyerges
Piotr Jankowski

THE GUILFORD PRESS
New York   London

© 2010 The Guilford Press
A Division of Guilford Publications, Inc.
72 Spring Street, New York, NY 10012
www.guilford.com

Printed in the United States of America

This book is printed on acid-free paper.

Last digit is print number: 9   8   7   6   5   4   3   2   1

Library of Congress Cataloging-in-Publication Data

Nyerges, Timothy L.
   Regional and urban GIS : a decision support approach / Timothy L. Nyerges, Piotr
Jankowski.
      p. cm.
   Includes bibliographical references and index.
   ISBN 978-1-60623-336-8 (hardcover : alk. paper)
   1. Geographic information systems.   2. Geospatial data.   3. Decision support
systems.   4. Decision making.   5. Regional planning—Decision making.   6. City
planning—Decision making.   I. Jankowski, Piotr.   II. Title.
   G70.212.N94 2010
   307.1′20285—dc22

                                                    2009028917

Preface

Geographic information systems (GIS) are being used by more people and organizations for more complex decision problems than ever before. Some of the most challenging decision problems for regional and urban communities involve how to plan, fund, and implement transportation improvements that are socially equitable, land developments that are ecologically sound, and water resource protection strategies that are economically viable. Such decision problems incorporate sustainability challenges in the sense that future generations would be better off if they were provided with better access to economic, social, and ecological resources.

Regional communities refer to ecologically diverse, large geographic areas affected by economic, social, and ecological impacts of human activities. For example, large areas contain transportation systems with many modes for movement, such as buses, trains, light rail, cars, and trucks, but these modes might not be available to all, which is a challenge to social equity. Another example might be watersheds containing residences for hundreds of thousands or even millions of people, all having some impact on land use conditions. Urban communities refer to the densely populated places wherein human and/or natural phenomena are likely to cause or be impacted by external effects from human or natural events; urban places are geographically smaller, but the human activity is more intense.

Although GIS software and hardware technology has matured substantially over the past three decades to address the problems mentioned above, there is still a tremendous opportunity to improve problem solving and decision making by enhancing people's "brainware" for gaining insight about the relationships among economic, social, and ecological concerns. To address complex decision problems using a sustainability approach, we add decision analysis and communication to already established data management, spatial analysis, and visualization capabilities for working with geographic information. Economic, social, and ecological objectives, and the trade-offs involved in pursuing them, are at the core of growth management perspectives. When we add issues dealing with intra- and intergenerational equity, then we deepen the information perspective into regional and urban sustainability.

Many researchers and practitioners have complained that sustainability as a process lacks a practical approach. To address this concern, *Regional and Urban GIS* offers readers a practical approach to using GIS when faced with regional and urban sustainability decision problems. We explain how to work within the pervasive planning, programming, and implementation decision situations facing regional and urban communities, particularly situations that involve land, transportation, and water resources.

The book is organized into 14 chapters contained in five parts. The chapters in Part I situate GIS within the decision support approach: Chapter 1 introduces the need for GIS in decision support. Chapter 2 provides a framework for various dimensions of decision problems. Chapter 3 introduces GIS workflow strategies that are used throughout the book to organize data processing activity. Chapter 4 shows how decision situation framing is a first step in coming to grips with decision complexity.

Part II includes chapters about a variety of GIS-based data analysis methods. Frameworks for different types of tools to address different types of geographic data play an important role in organizing choices about how to analyze geographic data. Chapter 5 describes the importance of developing databases to meet decision information needs. Chapter 6 presents several tables that organize a variety of data analysis techniques based on the types of spatial relationships and the types of data that are part of various problems. Chapter 7 introduces multiple criteria evaluation techniques that are the foundation of many decision analysis approaches.

Part III lays out how to embed decision analysis methods into planning, improvement programming, and project implementation decision situations. Chapters 8–11 present case studies about planning, improvement programming, and project implementation, incorporating land, transportation, and water resource decision activities. The case studies offer diverse perspectives about how to put the GIS tools to use with different kinds of data.

Part IV contains advanced topics about linking across data themes (like land, transportation, and water resources) and across decision processes (like planning, programming, and project implementation)—two important perspectives for implementing sustainability management. Chapter 12 introduces readers to linking across data themes, fostering an understanding about trade-offs in economic, social, and ecological concerns. Chapter 13 introduces readers to linking across decision processes, thereby establishing an information flow that connects long-term, medium-term, and short-term decision time frames.

Finally, Part V provides perspectives for using GIS that incorporate sustainability principles, thereby guiding the reader in how to frame the use of GIS for developing sustainable regional and urban communities.

Software applications appear throughout the text to provide readers with a hands-on perspective about GIS packages. Review questions appear at the end of each chapter, offering readers a means for self-study.

Early readers of this text have noted that complex decision situations are now easier to understand and more solvable with the aid of the GIS frameworks presented in the book. We hope all readers find this book as informative and useful in their GIS learning activity.

Acknowledgments

Many of the ideas and much of the material assembled for this book stem from our interactions with numerous colleagues in universities and local, state, and national agencies over the past decade. We gratefully acknowledge support for research and education that has generated many ideas, particularly funding from the National Science Foundation programs in Geography and Spatial Sciences together with Decision Risk and Management Science under Grants No. SBR-9411021 and No. SBR-0111481, as well as from the cross-foundation special program on Information Technology Research under Grant No. EIA 0325916. In addition, we acknowledge agency support from the project titled Consortium for Risk Evaluation with Stakeholder Participation, funded by the U.S. Department of Energy Cooperative Agreement DE-FC01-95EW55084, as well as project funding from the Idaho State Board of Education. The above support does not constitute an endorsement by the organizations of the views expressed in the book, but it helps to demonstrate that it takes diverse ideas to assemble a sustainability perspective.

Our perspectives have benefited from interactions with many colleagues. Some of these interactions have been with local and regional agency personnel, as in the City of Seattle, King County, and the Puget Sound Regional Council. We are grateful for collegial interactions with personnel from the Idaho Department of Water Resources, particularly with Director David Tuthill. Without his insights to address water resource planning in an analytic and deliberative way in Idaho, we might not have been as confident as we are that the kinds of GIS methods and technology described in this book are effective. Such methods really do work for complex and practical problems. We are indebted to our students for providing comments on extensive portions of this material, as it was being classroom tested over the past several years. A special thanks to our former and current graduate students at the University of Washington, University of Idaho, and San Diego State University. At the University of Washington, we thank Michalis Avraam, Jean Carmalt, David Moore, Kevin Ramsey, Guilan Weng, Matt Wislon, and Guirong Zhou. At the University of Idaho, we thank Steve Robischon and Alan Smith. At San Diego State University, we thank Arika Ligmann-Zielinska and Martin Swobodzinski. Their efforts in working with the material and with the students in our courses are truly appreciated.

We thank the staff at The Guilford Press, particularly our Editor, Kristal Hawkins, for seeing the value in a decision support approach to regional and urban GIS as a book, and Senior Production Editor, Anna Nelson, for working through the book production process with us. We also thank the many colleagues who contributed to their efforts.

Lastly, no acknowledgment of this nature would be complete without thanking our spouses for putting up with the moods that sometimes surface when working through particularly "sticky" challenges with idea development and draft rewrites. Thanks go to Pat and Ania for their endearing and continued support.

Contents

PART IV. Using GIS for Integrated Decision Support

PART V. Concluding Perspective

List of Tables, Figures, and Plates

TABLES

FIGURES

PLATES

PART I

Situating GIS-Based Decision Support

CHAPTER 1

Introduction

Need for Geographic Information Systems in Decision Support

This book is designed to provide a learning experience about geographic information system (GIS) use while focusing on a decision support approach within an urban regional context. *Decision support* refers to the tools and information provided by/to people during all aspects of their decision-making processes. The material is presented within the context of decision situations encompassing planning, capital improvement programming, and implementation for addressing land, transportation, and water resource concerns. Projects on the ground (or water, for that matter) exist in all three decision situations; that is, plans are commonly collections of projects. Improvement programs select from those projects and identify funds to implement some of them, and implementation results in realizing the project(s) as a result of their having been funded and developed. The decision situation concepts addressed in this book are applicable across a variety of thematic areas, such as social services improvement or ecological habitat rehabilitation, in addition to land, transportation, and water resources, but there is never enough room in a text for everything. Toward that end, this text applies concepts from *planning*, *programming*, and *implementation* (PPI) efforts to inform GIS use to various human–environment and environment–society decision situations, perhaps better called human–environment–society decision situations.

PPI activities are not the only pervasive decision situations that influence human–environment–society relationships. Communities sometimes make major investments, and now and again they face emergency management situations. Major investments, such as reintegrating very large tracts of land from military bases into the everyday use of communities, or regional transportation projects costing billions of dollars, or regional wastewater treatment facilities, each include all three PPI phases in a single project that might takes years, hence the "major" qualification. Furthermore, emergency management decision situations, for example, as in recovery efforts, also involve a time-

3

compressed version of the PPI phases. Consequently, these situations are very challenging to address. These latter two situations are not treated separately in this text. However, certain aspects of the three decision situations apply to each; thus, the material in this text is useful in understanding some of the needs in major investment decisions and emergency management situations.

We need to recognize that most decision efforts are commonly directed either separately or together by policy statements. Policy statements are formulated by organizational decision makers to establish a direction for work activity within the organization or across multiple organizations, as in public–private not-for-profit coalitions. A policy statement may in fact follow directly from law (e.g., growth management law) to construct a sense of societal behaviors in a place. For the most part, this book considers conventional approaches to community development, growth management, and newer ideas related to sustainability management as the focus of policy statements, without diving into the myriad details of why these policies are important. Nonetheless, it is important that the reader recognize the significance of such policy mandates for providing motivations for undertaking GIS-based decision support work.

We treat decision processes and substantive societal concerns within the context of GIS methods in an integrative way. This perspective leads us to issues about urban growth management in connection with community and regional sustainability management. Growth management and sustainability management are related in fundamental ways to integrative resource management, which we highlight later in this chapter.

GIS technology is fundamentally an integrative technology. As such, it is well-suited for addressing complex concerns that by their nature require an integrative approach to information development and use. Among the more intricate and important topics in the 21st century are group-based decision support efforts that address multifaceted infrastructural concerns and complex decision-making processes. Participatory approaches to resource management are on the rise in democratic nations all over the world.

This book is meant to foster a learning experience for GIS analysts who wish to take their GIS skills to the next level. However, this is more than a methodological treatment of software techniques. The substantive treatment of land, transportation, and water resources, plus the concepts underpinning decision support, provides GIS analysts with a well-balanced learning perspective.

The fundamental learning objectives for GIS analysts making use of this book are as follows:

- *Understand* the intellectual benefits and costs of integrated data processing strategies with GIS, particularly within the context of urban–regional growth management and sustainability issues, including (but not limited to) problem definition, database design, data collection, data structuring, data analysis, and information presentation.
- *Grasp* the significance of group work to facilitate broader and deeper understanding and valuation of the use of geographic information to address complex urban, geographic, and environmental issues within the context of a pluralistic society.

- *Master* several GIS data processing strategies that form the basis of GIS software as a tool for critical inquiry that reflects multivalued interests for community, urban and regional planning, programming, and implementation decision situations.

To accomplish those learning objectives we begin by situating GIS in three contexts that together foster a comprehensive understanding of GIS development and use.

1.1 Perspectives on GIS: A Decision Support Approach

The maturing of GIS can be described in terms of three realms of development: GISystems, GIScience, and GIServices.

1.1.1 GISystems

Developments in geographic information systems (GISystems) have proceeded for over 40 years (Chrisman 2005; Foresman 1998). From those developments a number of important perspectives have surfaced. Clarke (2003 pp. 2–6) lists several definitions that reflect a variety of GIS perspectives but fail to synthesize those perspectives cohesively. Chrisman (1999a) presents a definition based on how information is represented, without directly incorporating key themes of information technology. The following definition combines three perspectives—components, processes, and motivations—to shed light on the meaning of GIS.

The GIS working definition in this book is *a combination of hardware, software, data, people, procedures, and institutional arrangements for collecting, storing, manipulating, analyzing, and displaying information about spatially distributed phenomena for the purpose of inventory, decision making, and/or problem solving within operations, management, and strategic contexts as related to issues at hand.*

Three perspectives are drawn out by this definition in the following way (Nyerges 1993):

1. Components of a system: *a combination of hardware, software, data, people, procedures, and institutional arrangements.*
2. Processes utilized within the system: *for collecting, storing, manipulating, analyzing, and displaying information about spatially distributed phenomena.*
3. Motivations for system use: *for the purpose of inventory, decision making, and/or problem solving within operations, management, and strategic contexts as related to issues at hand.*

Cowen (1988) was among the first GISystems researchers to publish an article that defines GIS in terms of a decision support system. He recognized that all aspects of GIS in some way assist with decision making and effectively provide support in decision-making contexts. This purpose has been adopted as a popular way to describe GIS usage, no matter what kind of application one considers.

This textbook is the outcome of many years of work articulating GIS as applied to decision support systems. Although all of the components are important, the data and the software are the focus of this book. Educating people to become well-balanced "GIS analysts" is a major thrust of our efforts in this textbook. Consequently, we spend considerable time on these issues in chapters to come, but here we review some GIS basics by discussing GIScience and GIServices.

GIS data comes from a variety of sources addressing a variety of themes within and among organizations all over the world. Analysts integrate multiple sources of GIS data for a more holistic view of complex situations. Data integration occurs through software manipulation. The power of information in GIS stems from the way in which data management software technology supports complex spatial analysis software technology, which is in turn displayed in vibrant ways using map visualization software technology. For example, within the ArcGIS® product developed and distributed by Environmental Systems Research Institute (ESRI; which has the largest market share of GIS applications in the world) exist three basic modules. The most fundamental is ArcCatalog, which supports data management. Data management operations are core because GIS cannot operate without data. ArcToolbox contains spatial analysis software that works with the data in ArcCatalog. The ArcMap module can pull data from ArcCatalog or render the results of ArcToolbox visually. In all cases, each component interacts with the others to make GIS what it is. It is in this synergy that GIS takes a front seat as one of the foremost information technologies, emerging and evolving into a powerhouse of applications. Although GIS data and software techniques and methods are the focus of this text, the concepts behind GIS have been integral to GIS growth and maturation. Growth in GIS data and software development and use has been spurred by concomitant development of fundamental concepts that we refer to as GIScience.

1.1.2 GIScience

Georgraphic information science (GIScience) is the essence and foundation underlying a robust development and use of GISystems (Longley, Goodchild, Maguire, and Rhind 2005). A *science* can be defined as a systematic treatment of verifiable knowledge about the elements, structure, and process of reality. GIScience involves systematic treatment of conceptual information about topical (thematic) concerns that are described generally in terms of (geo)spatial, attribute, and temporal aspects of (some portion of) reality, and the interrelationships among those aspects. Systematic learning about GISystems and its underpinning in GIScience involves a mix of three knowledge domains: theoretical (conceptual), methodological, and substantive. When mixing these domains in different balance we arrive at different approaches to research investigation, as well as different foci of teaching and learning. A good book, as well a good intermediate course in which such a book is used, provides a reasonable balance among these three domains. A reasonable balance is one in which the presentation of material supports reader engagement through all three domains, although a reader might readily grasp material from one or two of those domains to start.

1.1.3 GIServices

Developments in Internet and mobile device technology are constantly emerging. Such technologies are referred to as *distributed GIS services*, as part of the title of a book about Internet GIS (Peng and Tsou 2003). Such services provide access to customizable applications—sometimes referred to as *geographic appliances* because of the personal-like flavor of utilities and their pervasive availability. Geographic information services (GIServices) effectively reflect the technology and innovation behind GISystems in everyday settings. Many cell phones have a built-in global positioning service for locating the person using the phone. Web services are a relatively new form of web-based systems development. Such developments will likely become increasingly important as many developments head in that direction.

All three contexts of GISystems, GIScience, and GIServices are receiving increased attention from people all over the world. The opportunities for important work abound. Nonetheless, a textbook cannot cover all elements of this developing field. We must choose a focus to make this text relevant, consistent, and meaningful, and, we hope, foster an engaging learning experience. Our focus in this textbook is on using GIS for decision support to address engaging topics.

1.1.4 GIS as Decision Support Systems

As mentioned previously, Cowen (1988) described GIS as a decision support system involving the integration of spatially referenced data in a problem-solving environment. Others, who argued that GIS technology fell short of providing decision analysis capabilities, quickly disputed this definition. Another perspective, which we offer later, argues that GIS might be a step backward due to its positivistic approach, encouraging rational planning in the decision process rather than opening the decision process to participatory behavior. Considerable progress has been made over the past two decades if we measure the progress in terms of tool development.

The basis of geospatial decision support is the GIS technology. The basic decision aids of GIS include data management to extend human memory, graphic display to enhance visualization, and spatial analysis functions to extend human computing performance. Beyond these common GIS decision aids, special features include modeling, optimization, and simulation functions required to generate, evaluate, and test the sensitivity of computed solutions. Other functions, such as statistical, spatial interaction, and location/allocation models, can be found in special GIS software packages. However, instead of expanding a GIS toolbox indefinitely by adding new models and procedures, software designers decided to open the toolbox up for modelers by providing *application programming interfaces* (API), which allow enhancement of the decision support function of GIS by adding models that support various capabilities. Examples of such decision-aiding models linked with GIS include various environmental models and multiple criteria decision making (MCDM) models used for evaluation of land planning decisions. Developers of special GIS decision support software—often called *spatial decision support systems* (SDSS)—have pursued various strategies of linking analytical models with GIS. They range from file exchange mechanisms (so-called *loose coupling*); data exchange protocols, such as dynamic data exchange (so-called *tight coupling*); all the way down to implementations of predictive/prescriptive models and decision support functions in GIS toolboxes (so-called *embedded coupling*). The possibilities of rapid SDSS development through linking analytical models with GIS have been recently expanded by various software technologies, including the open source programming language Python and the Sun Microsystems Enterprise JavaBeans technology. Approaches to object linking and embedding foster integrated applications support for data management and can only help with integration of data management, analysis techniques, and visual representation—the three core functions of SDSS. The special development of GIS into SDSS over the past several years has motivated commercial packages to expand as well. In this textbook we introduce a mix of decision support capabilities from both commercially available and special GIS packages.

At the same time as spatial decision support system development was moving forward, developments involving *planning support systems* (PSS) were getting under way (Shiffer 1995; Brail and Klosterman 2001). The focus was on how to make use of decision support capabilities incorporating GIS and analytic models in a planning context. Rich multimedia displays are a big part of that development, because the multimedia broaden the channels of communication for community participants (Shiffer 1995, 1998, 2002). PSS developments are related to the GIS-based activity, but PSS have been conducted mostly in the context of planning for groups (Geertman 2002a, 2002b; Geertman and Stillwell 2004). GIS for planning support is one of the decision situation contexts treated in a later section, and is distinguished from improvement programming and project implementation decision situations.

1.2 Decision Support in Land, Transportation, and Water Resource Management

Human–environment, environment–society, and human–human relationships grow more numerous and complex as more people live closer to each other spatially. For example, although coastal counties comprise only 17% of the U.S. contiguous land area, 53% of the U.S. population lives in these areas (Crosset, Culliton, Wiley, and Goodspeed 2004) and, of course, many more people enjoy them throughout the year as a vacation destination, often creating a complex system of human–environment–society relationships. Those relationships stem from a combination of social, economic, and ecological conditions and impacts in an urbanizing world. Because GIS technology has matured on multiple fronts over the past decades, we are able to consider more of the relevant conditions and impacts when deciding how to address these relationships. For example land, transportation, and water resource (LTWR) management activities commonly deal with social, economic, and ecological impacts as part of complex decision situations. Such management activities and the associated impacts are often undertaken as part of intergroup and interorganizational PPI decision work: work that occurs at multiple spatial and temporal scales.

Our principal motivation for focusing on PPI decision work within urban–regional settings is that such work in such settings is pervasive within communities around the world. GIS has matured sufficiently to handle complex urban, social, economic, and/or environmental problems, providing further motivation for casting GIS work in terms of a decision support approach.

Many people would agree that land, transportation, and water resources, and the relationships between and influences among them, are fundamental issues in an urban and regional context. Each of these substantive topics, whether taken separately or addressed together, can benefit from an integration of GIS and decision analysis. Land resource activity is the basis of human existence on a macro scale. Broadly taken, it influences social, economic, and environmental conditions that form the basis of a livable place. As societies grow more specialized, more people tend to "get around," exchanging goods and services or interacting to fulfill human needs. Transportation decisions about infrastructure development and transportation mobility involve many issues, including vehicle and fuel technological changes, road and vehicle operations improvements, and demand management. One of the most important natural resources needed for sustenance is freshwater, but a mix of saltwater (marine) and estuarine water is also very important, particularly in coastal areas. Ecosystems rely on some level of water to sustain human, animal, and plant life.

The terms *planning, programming, implementation, management*, and *decision making* show up in book titles, journal articles, reports, and conversations about what people do with their time and energy when trying to anticipate and address the needs of communities. This book uses the concept of *management* as the overarching term for which policy, planning, programming, implementation, major investment, and emergency are the finer-grain categories, because each has about it some management component. Each of the first five categories includes a process that is somewhat routine, involving work in an everyday world. In addition, we would be remiss if we did not consider the "emergencies" in life that pop up now and again. Unforeseen emergency circumstances occur from time to time, which prompt us to act. The emergency management decision situation is recognized as significant and important. As we mentioned earlier, policy, major investment, and emergency situations are not addressed in this book due to space limitations. Consequently, our focus is on PPI decision support situations, and particularly LTWR management, due to its routine and pervasive nature.

All three decision situations occur as a means of influencing change in the world (e.g., the change in land and transportation systems in a growth management context). Decision situations are motivated by, and are thus reproduced in, laws, regulations, and policies; hence, they establish operating procedures in public, private, and public–private organizations. We can refer to such organizations as "institutions" if we consider the influence of reproduction of work activity over the long term within particular places. Clearly, if "what seems to have worked" is no longer working, then (sub)organizations (in councils, legislatures, etc.) can change and/or (re)write laws and/or (sub)organizations (in executive branches of government through departments or agencies) can (re) write regulations and/or (sub)organizations (in judicial branches of government) can reinterpret laws. Such laws, regulations, or reinterpretations are the fundamental motivations for (mostly government) organizations to (re)direct human, financial, and/or natural resources in ways consistent with the extant political power.

The ways in which PPI has been conceptualized and practiced transform a community over time in various ad hoc, systematic, and/or sometimes chaotic ways. The relationship among functional themes—LTWR—and decision processes—PPI—constitute the basis of conventional growth management and sustainability management approaches to decision support situations outlined in detail in Chapter 2.

Conventional approaches to LTWR decision making is giving way to more integrated approaches. Two major types of integrated approaches can be identified. One is called a growth management approach, or "smart growth." In growth management, explicit connections between and among functional themes (e.g., LTWR) are recognized. Because growth management planning has gained significant momentum within regions across the world, economic and population growth are on a significant rise. In a second integrated approach called sustainability management, not only connections among functional themes but also significant connections among the planning, improvement programming, and project implementation are recognized. Sustainability management takes on multiple spatial and temporal scales of reference simultaneously. Many jurisdictions are exploring GIS use for growth management as a step toward sustainability management (i.e., managing economic, social, and environmental [ecosystem] conditions in a community). Sustainability management extends the growth management approach with a greater degree of integration. Farrell and Hart (1998 p. 4) define *sustainability* in terms of two perspectives. One deals with competing social, economic, and ecological objectives or priorities. The other addresses ecosystem limits or carrying capacities. Both perspectives involve intra- and intergenerational equity. In a sustainability perspective we look to the limits of the ecosystems in which our communities function to articulate physical constraints to growth (Farrell and Hart 1998). Farrell and Hart see those natural constraints and competing social and economic priorities as the two foundations of a sustainability approach, but they include with that a temporal perspective extending over multiple generations, rather than one or two, as is commonly the case. Whether we take a conventional regulatory approach, a growth management approach, or a sustainability management approach to improve community well-being, planning, programming, and project implementation, decision situations underlie much of our institutional efforts. As such, it will be important to address decision situations to understand the potential of a GIS-based decision support approach.

A final important point to recognize throughout the book is how various groups of people take part in decision processes. Decision-making groups are the foundation for societal change (Poole, Seibold, and McPhee 1985). Small-group and large-group decision making has become more visible over the past two decades, because decision transparency, particularly in public settings, is right next to decision accountability as one of the primary concerns of the general public, as well as special interest groups, in wanting to know how public money is being spent. GIS maps can help decisions become more visible and remind us of our moniker, "We map what we value, and we value what we map." ℘

1.3 Overview of the Book

This book provides readers with the following:

1. *Frameworks* for understanding and practicing GIS use in several application domains.
2. *Perspectives* on GIS development and maturation of GIS as a decision support technology, which is but one important ingredient in the mix of resources employed by people who strive to improve livability and quality of life in communities.
3. *Methods* for addressing complex decision situations that can be supported with the use of GIS technology.

The frameworks, perspectives, and methods involve theoretical, methodological, and substantive issues that help to organize the material in this book. We highlight a number of these issues in a chapter-by-chapter overview.

In Chapter 2, we outline the planning, programming, and project implementation decision situations that frame the substantive issues in this book. Public and private organizations participate in such situations around the world, with some situations more prevalent in some countries than in others. We further contextualize the types of decision situations using three topical themes: land (use) development, transportation, and water resources.

In Chapter 3, GIS project workflow is presented to provide a foundation for how to do GIS. We use an example from fictitious Green County to describe both basic and more advanced workflow.

In Chapter 4, we present a framework for documenting the many aspects of a decision situation, such as task purpose, technology needed, process to be used, and expected outcomes. The framework, called *decision situation assessment*, has been used in several everyday examples, and an overview of how it can be applied at four levels of detail is provided. This material provides a conceptual framework for the material presented in Chapters 2 and 3.

In Chapter 5, we focus on database development, clarifying the difference between data models and database models. *Data models* provide a framework for software designs that are used to guide the development of database models—the outcome of a database design. We describe a database design process for the case study about Green County wastewater facility planning.

In Chapter 6, we describe a framework for GIS data analysis as a basis for the chapters that follow it. Choices for GIS data analysis stem from what one intends to accomplish by generating information. However, these choices are based on the fundamental capabilities inherent in GIS data methods.

In Chapter 7, multicriteria evaluation techniques form the basis for interactive decision analysis. A case study provides the foundation of how to use such techniques for Green County decision problems.

In Chapter 8, we focus on a comparison of planning processes and the GIS data analysis that supports this type of decision process. We compare various planning processes to provide insight about alternative ways to proceed with planning analysis.

In Chapter 9, we present a case study about conjunctive water resource planning that took place in the Boise River basin of southwest Idaho. A conjunctive administration framework organizes how to consider surface water and groundwater as a joint resource.

In Chapter 10, our focus turns to improvement programming. We address the projects that are part of plans, whereby financial considerations come into play. Transportation improvement programming is the case study we use in this chapter to highlight the nuances within GIS data analysis.

In Chapter 11, the emphasis turns to a project implementation decision situation. Project implementation is more detailed than either planning or programming. Mitigation analyses are most common for a project implementation level of decision situation.

In Chapter 12, we describe integration across functional LTWR themes. This integrated perspective is the basis for growth management planning, programming, and project implementation.

In Chapter 13, we further the integration perspective and discuss integrating across decision situations. Because the three decision situations—planning, programming, and project implementation—commonly differ with respect to time horizon, we contend this type of integrated perspective leads to sustainability management.

In Chapter 14, we draw together frameworks presented throughout the book and offer conclusions about the current and future use of GIS for growth management and sustainability management.

1.4 Summary

This book supports a learning experience about GIS methods using a decision support approach. *Decision support* refers to the tools and information provided to people during all aspects of their decision-making processes. The context for urban–regional decision-making processes in this book are planning, capital improvement programming, and project-level implementation decision support situations that address LTWR concerns. However, the concepts and methods presented herein are applicable across a variety of other thematic areas, such as social services improvement or ecological habitat rehabilitation.

Over the past decade, GIS concepts have matured into a GIScience that is the basis for a continued redevelopment of GISystems tools. Some of the tools are now thought of as everyday appliances; hence, the name GIServices has been coined to indicate the pervasive character of geographic information that is Internet-enabled. In this book we focus on the *systems* area, but knowledge about systems is clearly related to science and services, and we should not forget this. The definition of GISystems presented in this book highlights three perspectives: (1) components as the building blocks, (2) processes utilized in the system, and (3) motivations for system use.

Decision problems about urban–regional environments are among the most vexing problems facing communities worldwide, as the world becomes more urban. Because the urbanization trend will continue for quite some time, many people agree that land, transportation, and water resources, and the relationships between and influences among them, are and will continue to be some of the most complex issues in an urban and regional context. Each of those substantive topics, whether taken separately or addressed together, can benefit from an integration of GIS and decision analysis. Setting the GIS methods presented in this book within PPI decision support situations, particularly addressing LTWR decision problems, provides a solid foundation for a decision support approach to the use of geographic information.

The decision support approach in this book lays out a way to move from conventional management of urban–regional information to a growth management perspective, and on to a sustainability management perspective. That move entails broad-based participation in decision processes, an important part of enhancing pluralistic democratic processes. Several authors now describe *the public* as comprising several categories of diverse publics, because of pluralistic values sought in planning, programming, and implementation decision processes. How broad-based should such participation be to enhance the local knowledge to foster breadth and depth of perspectives, without overburdening the decision processes, is a matter of choice for communities.

1.5 Review Questions

1. What are the three perspectives for defining GISystems?

2. What is the advantage of using these three perspectives?

3. Describe the relationship among GIScience, GISystems, and GIServices. Why are these knowledge areas worth learning about?

4. Characterize your interpretation of a decision support approach using GIS. Why is a decision support approach to using GISystems useful in this day and age?

5. How does planning differ from improvement programming from program implementation?

6. Why are land resources, transportation resources, and water resources the substantive focus of the decision processes addressed in this GIS textbook?

7. How would you characterize the differences among PPI decision processes in terms of spatial and temporal scales?

8. How do conventional, growth management, and sustainability approaches to resource management differ?

9. Why is it useful to know about the groups of people participating in decision processes?

10. How does consideration of three decision situations—planning, improvement programming, and implementation—form the basis for understanding an integrative approach to using GIS?

CHAPTER 2

GIS in Decision Support Situations

In this second chapter we illuminate the challenges of urban–regional decision support situations. Making decisions in urban–regional communities is not new, because there are many conventional opportunities for making decisions, such as functional planning and zoning ordinance coordination. GIS uses abound in these situations. To move beyond these somewhat limited perspectives, at least a dozen U.S. states and many more communities within the United States and around the world have adopted some form of growth management regulations; other communities around the world have adopted sustainability management perspectives. Growth management attempts to tackle problems that arise in growing communities, for example, in most cases, population growth. GIS use is a mainstay in all of these communities as they examine the land use, transportation, and critical resource concerns facing them. Some states foster a top-down approach to management (control from the state level), whereas other states foster a bottom-up approach (control from the local level). Growth management approaches are compared and contrasted with a sustainability management approach. Threshold levels of growth are introduced in a sustainability management perspective, and they commonly take a longer-term perspective about community change. As part of a sustainability management approach we describe integrated perspectives on planning, improvement programming, and project implementation decision support. The five dimensions for integrating planning, programming, and implementation are organized into a framework for integrated, situation assessment. Making the situation assessment operational relies upon GIS being available. A summary provides the highlights of the chapter material.

2.1 Conventional Approaches to Decision Support Situations

Land use regulation has been in place for a while to address community change. Five main techniques serve as policy instruments for implementing a conventional regulatory approach to address community growth.

1. Community plans: 10- to 20-year horizon, multiple scales and foci.
2. Subdivision regulations and plans: developer plans required when land is subdivided.
 3. Capital improvement programs: infrastructures to serve the public (e.g., streets, parks, waterways, public buildings).
4. Zoning ordinances: the most common regulatory instrument.
5. Public participation: collecting feedback from people about their concerns.

Land, transportation, and water resources, among many other aspects of community well-being, are not new. Communities across the world have been managing these varied issues through conventional approaches to decision making. Plans (e.g., land use plans), capital improvement programs (e.g., transportation improvement programs), and project implementation ordinances (e.g., zoning laws) have been used for many years (Porter 1997). Plans encourage a look forward to address external influences of human activity at broad spatial and temporal scales. Capital improvement programs budget investments to correct for insufficient service conditions, thus providing a package of projects on a medium spatial and temporal scale. Project implementation focuses on project activity in the near term and over smaller spaces. We describe these techniques in the context of the three decision support situations—planning, programming, and implementation—that frame much of the material in this book. It is important to set the stage for how most communities address change. After all, not all communities are growing so rapidly that they actually need such advanced approaches to growth and/or sustainability management.

2.1.1 Planning-Focused Decision Situations

Regardless of the different conventional, growth management, and sustainability approaches, an inherent commonality to many approaches is that they all attempt a "general scoping" of issues. Because use of GIS is so pervasive across cities, counties, and regions, the result is often a "map-based vision" of what a future could be in regard to land use, transportation, and/or water resource functionality for those jurisdictions. This map-based plan tends to be jurisdiction-wide (e.g., for the entire city, county or region; see Plate 2.1 for a land use plan map and Figure 2.1 for a transportation plan map), but there are small areas for which plans are made as well.

Several steps are required to develop the land use plan for Middleton Township near Madison, Wisconsin. Start with a current base map containing the urban service area, tax parcels, and municipal boundaries. Added to the base map tax parcels would be the existing land composed of residential, public, parks and recreation, agricultural, industrial, and commercial uses. The current land use map provides a baseline for the growth projections. A zoning map is used to estimate where changes in land use might occur over the next several years depending on rezoned parcels. The zoning map establishes permitted land uses for particular areas. Population and employment projections computed outside of a GIS environment are commonly used to identify the expected extent

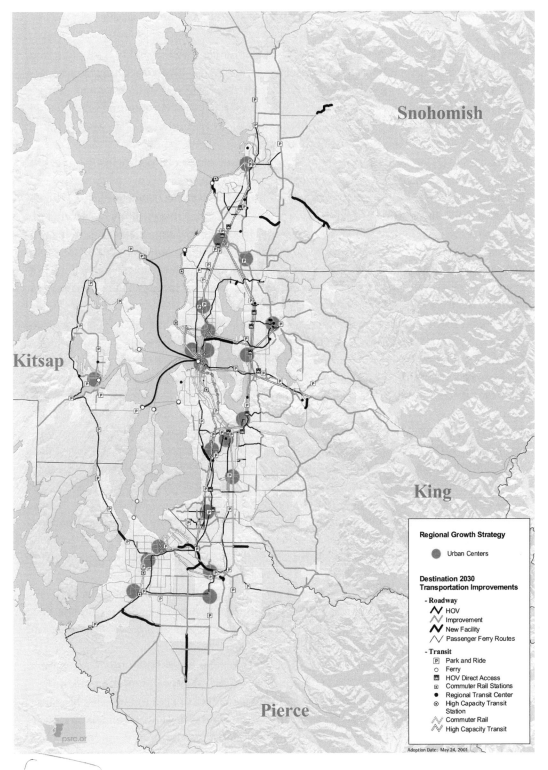

FIGURE 2.1. Destination 2030—Metropolitan Transportation Plan created by the Puget Sound Regional Council. From Puget Sound Regional Council (2001). Copyright 2001 by Puget Sound Regional Council. Reprinted by permission.

17

of change. However, data about existing land use indicate "where" the population and employment growth might situate themselves. Added to this, the agricultural productivity and sensitive natural areas are needed to identify areas that should be protected from encroachment by residential, commercial, and industrial uses, the latter of which would damage the overall ecological health of the entire area. We address such issues through database design and data analysis methods later in the book.

A land use plan helps government agencies create transportation plans. Changes in land use foster changes in transportation needs. Thus, future land use generates a need for future transportation infrastructure. The number of future trips generated by each transportation mode (i.e., pedestrian, bicycle, car, bus, train, and ferry) is estimated. Trips are then assigned to transport infrastructures to identify deficiencies. GIS is used in much of this work. However, transportation infrastructure changes foster land use changes. Thus, there is a mutual growth-encouraging relationship between land use and transportation, which is why they are treated together in a growth management context—but more about that later.

Decisions that enter into a planning vision are typically for the long-term, say, on the order of 10–20 years. However, short-term planning studies can also be undertaken. The difference commonly relates to the size of the area under consideration. Long-term plans commonly address large areas, on the order of a city, whereas, short-term plans address subareas. Kelly and Becker (2000) point to numerous decision points that exist within a variety of community planning contexts, both long- and short-term, for comprehensive plan use (e.g., land use, infrastructure, and water resource change).

Characterization of potential changes in so-called *rights of land use* is what makes area planning documents important in the planning process. If the use of land is not much different from one time period to another, then most people do not care about making land use plans, because the change is so minimal that "new map representations" are not useful. Drastic change is what concerns most people. Controlling changes, particularly long-term and/or numerous changes, calls for planning.

Randolph (2004) has extensively explored how land use planning supports environmental management that addresses the potential for change. He presents a variety of ways in which land use impacts human environmental health, hydrological systems, ecological resources, energy and material consumption, cultural heritage and community character, and environmental justice. These impacts stem from external effects of a land use activity on both the specific site of a land parcel and surrounding parcels. To address such impacts, Randolph presents a framework that differentiates a variety of plans, including long-range general planning, district planning, functional planning, and implementation plans. Long-range planning is the basis of comprehensive planning presented throughout this text. Comprehensive plans often take into consideration the influences of land use on transportation, as well as the influence of transportation on land use, but not always. The mutual reinforcing influences are always considered when planning under growth management regulations. However, comprehensive plans are often created without the mandate of growth management regulations. District planning covers a small area, rather than a citywide area, as in comprehensive planning. Functional planning focuses on individual themes (e.g., land use, transportation, and water resources) as described earlier. Implementation plans focus on specific project development and can for the most part be considered the same as implementation decision processes.

It is important not to confuse planning, improvement programming, and project implementation. The plan is a *general scoping document*, not a budgeting or operational representation of what should be undertaken. When money and/or other significant allocations of resources are involved, the decision process becomes an "improvement programming" decision situation. Implementation is about putting a project into action. However, to put a project into action, a community must budget for implementation. Improvement programming prioritizes projects to be budgeted.

2.1.2 Improvement Programming-Focused Decision Situations

An improvement programming decision situation considers what to change and estimates how much it will cost to make the improvement. As such, financing is an important part of the decision situation, in addition to all of the other factors considered in the plan. Consequently, an improvement program is a collection of projects, with each project having associated benefits and costs—but not necessarily expressed in monetary terms. Unfortunately, seldom do communities estimate the cumulative benefits and costs associated with various *packages* of projects. Each package is associated with a scenario of assumptions (e.g., financing conditions), such as ways to raise money through user fees or excise taxes, or some other mechanism to pay for the entire package of projects. That package, together with the scenario of assumptions, comprises the improvement program.

An improvement program decision situation provides a more refined perspective on land use, transportation, and/or water resource change than does a planning decision situation. The projects are clarified a bit more when "dollars and cents" are involved in an improvement program. However, because the focus is on potential projects and potential financing, a detailed assessment of the impacts (i.e., benefits and costs) is not part of this process. There are seldom enough resources to dedicate to this level of decision situation as it deals with collections of projects. This is particularly the case in terms of social and environmental assessments, even if the economic assessment is part of the process.

Hillsborough County, Florida (i.e., the county that includes Tampa), provides a website for capital improvement programs that makes extensive use of GIS map displays (Hillsborough County 2006). Plate 2.2 shows the web page for the project information management system. That system helps county employees, as well as the general public, track projects for all capital improvement programs, including project categories for the following: fire services, government facilities, library services, parks, potable water, reclaimed water, solid waste, storm water, transportation, and wastewater. At last count, the system was tracking 445 projects.

2.1.3 Implementation-Focused Decision Situations

Zoning ordinances regulate land use projects such that property owners exercise their rights so as not to injure each other (see Plate 2.3); *land use actions* are land projects that occur, based on zoning laws. Houston, Texas, is the only major city in the United States that does not have zoning laws; regulation is by deed and community, covenants, and restrictions. Zoning laws were first established in New York City in the early 20th century to reduce the external effects of industrial land use on residential land use. Property owners are required to make changes within the constraints of the land use regulations.

A project implementation decision situation is one in which a detailed economic and/or social, and/or environmental assessment is performed. At this level of decision situation, individuals (planners or consultants, etc.) who are responsible for the detailed analysis only have to consider the impacts related to a specific project. Given that the focus is narrow (in space and time), much more energy can be spent examining details of how a particular project might impact a community. The project analysis might actually include several alternatives for a project. However, little, if any, cumulative impacts can be examined in this situation, because a single project is commonly the focus. Information from other projects is usually not considered; the law does not commonly require it.

Developers (whether commercial or private individuals) commonly subdivide large land parcels into smaller land parcels to make more intense use of the land—hence, the term *subdivision*. Subdivision regulations, and the documents that address those regulations, describe how the land can/will be divided (Brown 1989). Subdivision regulations are the laws that guide the subdividing process in accordance with current zoning laws. A map of a parcel of land that is subdivided for use is called a *plat map* (Figure 2.2). *Subdivision plats* are documents that many communities use to address the proposals for land use change at microscale detail, although, clearly, large projects covering multiple land parcels are possible.

In a subdivision of land, the *right to use* land is transferred from entity to entity (i.e., person to person, organization to person, person to organization) in the form of property rights. The right is transferred as a *title* to property. This entitles the owner to exercise certain property rights as specified within the title. Thus, regulations to use land might be set by the community and/or by transferring the title from entity to entity.

In conventional approaches to planning, improvement programming, and project implementation analysis, the decision situations are separate. Assessments are performed by different people at different times, and often in different units of an organization. In areas where there has been considerable population change, particularly in the last 30 years, the conventional approach to decision situations has been insufficient to address intensifying land use, transportation, and water resource change.

Part of the awareness about the inadequacy of standard regulatory approaches to land use change has come from *public participation*, which provides a mechanism for

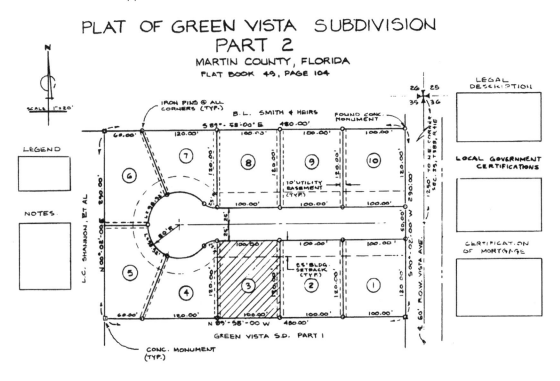

FIGURE 2.2. A plat map. From Brown (1989).

agencies to gain feedback from people about the institutional processes under way. Hopkins (2001) suggests five benefits of participation:

- Participation of more persons and more diverse persons increases group capabilities to make plans.
- Participation of decision makers increases the likelihood that they will use the plan.
- Participation of all constituencies avoids later resistance to chosen actions.
- Participation outside of formal democratic processes complements these processes by giving different people access and, consequently, representation.
- The experience of participating helps to foster the kinds of individuals necessary to operate a democracy.

Public groups serve to provide input, feedback, and recommendations about what governing bodies should do, which fosters collective choice in a democracy. The growth in stakeholder, group-centered participation has been spurred on by environmental laws, specifically recognizing that such participation fosters a democratic approach to decision making (Randolph 2004). The primary rationale for enhanced stakeholder participation in public land planning is based on the democratic maxim that those affected by a decision should participate directly in the decision-making process (Smith 1982; Parenteau 1988). Agencies, that is, elected officials and technical specialists within agencies, have been criticized for not having sufficiently broad perspectives to interpret and to understand all impacts that might arise within planning, improvement programming, and project implementation decision processes. Therefore, public participation interests have blossomed over the past 20 years to bring balance to decision processes from interested and affected parties. Recognizing this opportunity, many GIS researchers have been encouraging community groups to adopt GIS, and community groups have been asking for access to GIS to address a variety of community decision concerns (Craig, Harris, and Weiner 2002).

Some GIS researchers have begun using updated versions of the ladder of participation developed by Arnstein (1969). One update that is gaining favor is a participation spectrum outlined by the International Association of Public Participation (2005), as presented in Table 2.1. We can add to that some of the important activities that differentiate one level from the next. Implementing these levels in any participatory process essentially characterizes the difference between a weak and a strong democracy (Barber 1984). Informing people is only the very beginning of participation. If that is the extent of participatory process (like reading information from a website), then we can say we achieved at least a "weak" sense of democracy. If however, the public is empowered to take part in the decision process, then we can say we have achieved a "strong democracy."

Although more and more communities are fostering participation in decision processes, most are still very far from reaching a strong democracy. Participation requires investments in both time and money, so most communities at most reach an *involvement* level, with little information technology support. When geographic information technologies are introduced, the novelty of the process tends to foster higher levels of participation. Participation processes appear to be expanding in frequency and num-

TABLE 2.1. Participation Spectrum, Activities, and Impacts

Participation level	Participation activities	Public impact on overall process
Inform	Listen	Public is informed
Consult	Listen, respond	Public is informed and provides feedback
Involve	Listen, respond, negotiate, recommend	Public concerns are incorporated
Collaborate	Listen, respond, negotiate, recommend, analyze	Public helps form concerns and solutions
Empower	Listen, respond, negotiate, recommend, analyze, decide	Public helps decide concerns and solutions

ber of people involved (Gastil and Levine 2005). A significant amount of participation underpins the success of growth management processes, which makes such processes different than standard regulatory processes.

2.2 Growth Management around the United States

How is a growth management approach different from a standard regulatory approach to addressing community change? In a growth management context, a community identifies the *growth problem* as a specific category of concern. A *growth management approach* is a way of organizing community efforts to anticipate future development and problems that might occur. Consequently, a growth management approach, and techniques to implement that approach, stem from concerns about community problems over time (i.e., change in the community), for example:

- Managing the location and character of community expansion (e.g., urban growth boundary, development policy area, infill-redevelopment).
- Preserving natural resources and environmental qualities and features (e.g., land acquisition, conservation zoning, water quality/erosion control regulations, delineating critical areas).
- Ensuring efficient provision of community infrastructure (e.g., functional infrastructure plans, facility exaction, impact fees, transportation demand management).
- Maintaining or creating desirable quality of community life (e.g., design reviews, incentive and performance zoning [bonuses for mixed use and density], historic ➤ preservation).
- Improving economic opportunities and social equity (e.g., economic development incentives, affordable housing programs).
- Regional and state guidance of community development (e.g., coordination of local planning, development review with a regional impact).

Because these problems describe a type of community change, a GIS as an inventory of phenomena across space and time is one way of representing a basic understanding of that change. Change in land use activity, as in housing and commercial development, and/or transportation activity, as in the mobility of freight and people, and/or water resource activity, as in the degradation of waterways, are only a sample of growth management concerns that can be addressed by a set of thematic maps for various time periods.

In the United States, 11 states have enacted growth management laws, and others are under consideration (Pope 1999). Three states—Florida, New Jersey, and Oregon— have been using top-down controls (i.e., a strong, state-level control to encourage development growth). Eight states—Georgia, Hawaii, Maine, Maryland, Minnesota, Rhode

Island, Vermont, and Washington—use bottom-up control (i.e., stronger local-level control). A 12th state—California—is beginning to use a combination of both. Twenty-seven states have a role in growth management, but 13 have no mandated state laws.

In *top-down planning* states, such as Oregon, goals are more specific at the state level than they are in *bottom-up planning* states, such as Washington. In top-down states, the goals are stated in such a way that all counties within the state plan in the same way. In bottom-up planning states, the goals are generalized but made specific by local jurisdiction implementation. Certain development thresholds can differ from jurisdiction to jurisdiction, for example, from county to county or city to city.

Regardless of whether states use a top-down or a bottom-up approach, there are six types of intergovernmental planning responsibilities.

1. State plans (e.g., Florida, New Jersey, and Oregon use such plans as a top-down approach).
2. State agency planning and coordination—all states engage, top-down by mandate, bottom-up through a funding penalty (i.e., monies are not transferred to local jurisdictions from state to the local jurisdiction unless plans are completed satisfactorily).
3. Requirements for local planning—all states require certain portions of the state to undertake planning (e.g., fastest growing counties by population percentage).
4. Provisions for regional coordination—all states require some form of regional coordination (e.g., across multiple jurisdictions, such as counties within a region).
5. Processes for achieving consistency between local and regional or local and state plans (e.g., consistency between city and county plans, and then county to regional plans).
6. Appeals or conflict resolution procedures (e.g., Growth Management Hearing Boards adjudicate complaints from community organizations, businesses, and citizens).

Tensions among state, regional, and local growth management are played out daily in the news. It is part of the political fabric of a community and can change from legislative election to legislative election. Negotiating action over social, economic, and environmental values as related to extant conditions is a major part of politics, and fundamental to making progress in urban–regional sustainability.

In Washington State, "currently, 29 counties and 218 cities (representing 95 percent of the State's population) are fully planning under the [Growth Management Act or] GMA. Ten counties and their cities are planning for resource lands and critical areas only" (Washington State 2006 p. 1). In 2002, King County, the most populous among all 39 counties in the state, had 47 cities of which Seattle was the largest. Not all planned under growth management strategies, because some are as small as 1,500 residents; but the majority did plan using growth management guidelines. To implement a comprehensive plan and organize access to information about growth management, the Washington State GMA (1990), specifies that a comprehensive plan can be a set of maps and/or a GIS. Indeed, the term *geographic information system* is cited in the law.

In the central Puget Sound region, the growth management strategy takes form as part of VISION 2020 (Puget Sound Regional Council 2006a). VISION 2020 is the land use element of the comprehensive plan. Destination 2030 is the transportation element of the comprehensive plan (Puget Sound Regional Council 2006b). The goals for such plans—hence, the maps that are expressions of those plans—are presented in Table 2.2.

Articulating goals in a growth management plan is a matter of addressing laws, regulations, and policies at multiple government scales. For example, the growth management elements for the City of Seattle Comprehensive Plan depend on the Washington State Growth Management Act, King County Policies, and City of Seattle Policies. The plan, "Toward a Sustainable Seattle," is a 20-year policy plan designed to articulate a vision of how Seattle will grow in ways that sustain its citizens' values. The plan makes basic policy choices and provides a flexible framework for adapting to real conditions over time. The initial building blocks of the Comprehensive Plan are the *elements* required by the state's GMA: land use, transportation, housing, capital facilities, and utilities. King County's Countywide Planning Policies require the addition of an economic development element, and Seattle Framework Policies (Resolution 28535) inspired the inclusion of neighborhood planning and human development elements.

TABLE 2.2. Summary of Washington State Growth Management Act Goals

- Encourage development in urban areas where public facilities and services exist or can be efficiently provided.
- Reduce urban sprawl.
- Encourage efficient, multimodal transportation systems.
- Provide affordable housing for citizens of all income levels, promote a variety of housing densities and types, and preserve the existing housing stock.
- Promote economic opportunity consistent with the capacities of the state's natural resources and public services and facilities.
- Respect private property rights.
- Provide timely, fair, and predictable permit review processes.
- Conserve and enhance natural resources.
- Retain open space, conserve fish and wildlife habitat, increase access to natural resource lands and water, and provide recreational opportunities.
- Protect the environment and enhance the state's high quality of life.
- Encourage citizen participation in the planning process and ensure coordination among jurisdictions.
- Ensure that public facilities and services are adequate.
- Preserve historic and archaeological resources.

Note. From Washington State (2009).

The ideas in the plan were developed over 5 years through discussion and debate, and the creative thinking of thousands of Seattle citizens working with city staff and elected officials. GIS was used by neighborhood groups to formulate those plans.

In Alachua County Florida, GIS services are being made available to all citizens through the GIS Division of the Department of Growth Management, under the Board of County Commissioners (Alachua County, Florida 2008). The Division operates in the following four areas:

- Geospatial decision support in various areas of urban–rural planning applications.
- Asset and record management for land administration systems.
- Web development and maintenance for Internet and Intranet.
- Conception, design, implementation, and enhancement of e-services.

GIS services and products for growth management are being made available from a web portal conceived, developed, and implemented entirely within the GIS Division of the department. The County provides a wide range of applications, available to the public for a fee (see Table 2.3).

New Jersey is another growth management state. It is the most densely populated and developed U.S. state, in which 20% of the land (about 1 million acres) is publicly owned open space and preserved farmland, of which forests comprise nearly 45%, both public and private. However, in communities all across the Garden State, once wide-open spaces now sprout sprawl according to the New Jersey Department of Environmental Protection (NJDEP). To address that problem, the governor enacted regulatory policy for awarding land use permits called Blueprint for Intelligent Growth (BIG). NJDEP is implementing the BIG policy using a map with green (move ahead permitting), yellow (cautious permitting), and red (no permitting) symbolism; hence, they call it the BIG map (New Jersey Department of Environmental Protection 2006).

The BIG map is intended as a decision support tool to reduce overdevelopment and congestion. Ultimately, the approach is intended to strengthen environmental and natural resource protection, and to promote development and redevelopment in areas that are appropriate from an environmental and planning perspective, by providing a common platform among agencies to identify areas that are appropriate for growth and areas that need stronger regulatory protection. The map is being incorporated into the State Development and Redevelopment Plan (the State Plan) map. In 2003, following a period of informal consultation with municipal officials and the public, the map was undergoing review prior to being promulgated as a rule-making proposal.

The examples of GIS use for growth management planning indicate that many jurisdictions are exploring the relationship between growth management planning and sustainability planning (i.e., economic, social, and environmental [ecosystem] conditions) in a community. Sustainability planning perspectives have been maturing over the past couple of decades. The World Commission on Development and Environment

TABLE 2.3. Sample of GIS Growth Management Web Services Accessible to Public

Interactive GIS applications

Our GeoGM Mapper allows for creation of custom maps offering access to 50+ GIS layers. GeoGM Searches can be performed based on Address, Tax Parcel number, Tax Parcel owner's name, and Section–Township–Range (STR).

Our Map Atlas, searchable by Section–Township–Range (STR), offers ready made pdf Maps for the one-mile area defined by Section–Township–Range (STR) or land grant. For each Section–Township–Range (STR) one can view and download ready made standardized maps of up to date Parcels overlayed with Zoning, Future Land Use, Wetlands/Floodplains, Strategic Ecosystems, 2 ft Topographic Contours, or Aerial Photographs.

Our Multimedia & GIS for Historic Structures in Alachua County integrates in an interactive GIS application, 960+ Florida site files, photographs, video clips with a voice narrative, detailed descriptive information, GIS layers, and much more. Searches can be done from the map or from the database.

Our Ecosystems Interactive Mapper allows one to view and explore the results of the LEMAC model, Alachua County's Decision Support System for landscape evaluation and characterization. A parcel search and other conservation geospatial layers are included. This Mapper is part of a specialized website we have developed on ecosystems studies.

An Interactive Map of the world, part of the GISCorps website we have developed, shows locations of GISCorps volunteers and missions. The Mapper looks live into the main GISCorps database and it updates as new volunteers sign on.

Our Interactive Map of Florida Counties helps you find web addresses for Florida counties and county seats. It takes a bit to load though.

Tracking building permits and zoning applications

Track by Application Number the status and history of Zoning, Zoning Variances, and Building Permit applications. Information goes back ~20 years. Results are dynamically integrated with corresponding maps in the Map Atlas.

Track by Commission Meeting Date the status and history of Zoning and Zoning Variance applications. Information goes back ~20 years. Results are dynamically integrated with corresponding maps, reports, and agendas.

Track by Section–Township–Range (STR) the status and history of Zoning, Zoning Variances, and Building Permit applications. Information goes back ~20 years.

Map gallery for view and download

Our Collection of Static Poster Maps. View, download from our collection of 25+ standard maps and their corresponding documentation (i.e., metadata).

Our Collection of Static Comprehensive Plan Maps. View, download the entire collection of 55+ maps of the newly adopted Comprehensive Plan.

Note. From Alachua County (2008). Copyright 2008 by Alachua County. Reprinted by permission.

(1987 p. 8) published in *Our Common Future* that *sustainable development* is defined as the ability of humanity "to ensure that it [development] meets the needs of the present without compromising the ability of future generations to meet their own needs." The Task Committee on Sustainability Criteria (1998 p. 36) explains in *Sustainability Criteria for Water Resource Systems* that

> sustainability is an integrating process. It encompasses technology, ecology, and the social and political infrastructure of society. It is probably not a state that may ever be reached completely. But is it one for which we should continually strive. And while it may never be possible with certainty to identify what is sustainable and what is not, it is possible to develop some measures that permit one to compare the performances of alternative systems with respect to sustainability.

The National Research Council (1999 p. 23) published in *Our Common Journey: A Transition Toward Sustainability* that

> while many definitions about sustainable development have appeared, each sharing a common concern for the fate of the earth, proponents of sustainable development differ in their emphases on (1) what is to be sustained, (2) what is to be developed, (3) the types of links that should hold between the entities to be sustained and the entities to be developed, and (4) the extent of the future envisioned.

These passages demonstrate the maturation of perspectives over time. It appears that the challenge for sustainability grows ever more complex. Whereas some organizations have thought that sustainability (or striving toward such) is beyond what society can accomplish at this time, many organizations nonetheless encourage a movement toward sustainability even in the current economic times. For example, the mission statement of the Washington State Department of Ecology (2006 p. 9) "is to protect, preserve, and enhance Washington's environment, and promote the wise management of our air, land, and water for the benefit of current and future generations." This statement is consistent with the growth management laws currently in place, and it goes beyond them toward sustainability. To understand the connection between growth management and sustainability, a framework that describes characteristics of each perspective may be useful and help to coordinate goals.

2.3 Comparing Growth Management and Sustainability Management

It is advantageous to relate community and regional sustainability to growth management, if community and regional sustainability is to make progress within current institutional contexts. Drawing growth management and sustainability views into focus, we recommend a *community and regional sustainability* perspective that utilizes Farrell and Hart's (1998) competing social, economic, and environmental objectives for communities, which may or may not be considered alongside carrying capacities, and Rees's (1998) description of the importance of generational equity in sustainable community development (see Figure 2.3).

Competing objectives, carrying capacity, and intra- and intergenerational equity combine to form a progression of weak-to-strong community and regional sustainability. Growth management concerns are about competing objectives and intra- and intergenerational equity (weak and semi-strong sustainability), but growth management seldom addresses social, economic, and environmental concerns simultaneously. The natural, physical, and social sciences continue to assess carrying capacity related to various social, economic, and environmental contexts, the basis of "integrated assessment science" and considered the core of "sustainability science" (Kates et al. 2001). Sustainability assessments cut across jurisdictional boundaries (e.g., as in watershed sustainability studies). Watersheds do not align themselves nicely with political governance—and the problems of sustainability do not either.

Sustainability management research has led to articulation of approaches to sustainable cities (Haughton and Hunter 1994) and ecological cities (White 2002). Although the terminology is slightly different, the intent is the same for improving city well-being. Social, economic, and ecological developments must be in balance if cities are to move into the 21st century as places of overall well-being.

Insight into land, energy, transportation, and environmental health relationships presented by White (2002) provides a foundation for a *systems process model* of ecological cities. GIS can implement *dynamic process models* only in terms of increments

	Generational Equity	
	Intragenerational perspective	Intra- and intergenerational equity perspectives
Social, Economic, and Environmental Objectives and Constraints		
Competing objectives considered	*Weak sustainability* as a concern about competing objectives from an intragenerational perspective	*Semi-strong sustainability* as a concern about competing objectives from intra- and intergenerational perspectives
	commonly considered growth management	transition from growth to sustainability management
Competing objectives and carrying capacity constraints considered	*Semi-strong sustainability* as a concern about competing objectives and carrying capacity constraints from an intragenerational perspective	*Strong sustainability* as a concern about competing objectives and carrying capacity constraints from intra- and intergenerational perspectives
	transition from growth to sustainability management	sustainability management

FIGURE 2.3. A framework for characterizing community and regional sustainability in terms of three levels: weak, semi-strong, and strong. Weak and semi-strong sustainability can be considered growth management in some circumstances.

of a process (i.e., investigating what is related connect to what in a structural manner between two time periods), without resorting to other specialized software or considerable software scripting/programming within GIS. The effort and advantages of using a map (actually, a collection of maps) to depict the structural relationships of a process have been more useful to more people in small and large communities alike than rendering that same spatial process in computers. The evidence comes from the sheer growth of GIS implementation through systems process modeling in comparison to dynamic process models within towns, cities, counties, regional organizations, and states. One of the contributing factors is the integrative breadth of GIS technology versus the specialized focus of dynamic systems models. Thus, when "push comes to shove" in a budget for implementing information technology, GIS has won almost every time. However, we should make it clear that good solutions for complex problem solving actually require both kinds of technology. Steinitz and his colleagues (2003) have shown this time and again for real projects: The structural representation models and process representation models are both very important; neither replaces the other. More will be said about this in Chapter 3, when we introduce "task analysis" about wastewater facility siting.

Cities do not exist within and of themselves; they always have a hinterland, also called the region. Because cities exist within a regional context, and transportation is one of the biggest contributors to greenhouse gas emissions, regional transportation systems must be a focus of the greenhouse gas reduction effort within ecological city regions. Although individual cities (e.g., Seattle) and the resources used therein are very important to the overall effort of reduction, all cities within a functional region should be a focus of the greenhouse gas reduction effort, because the surface transportation is a regional infrastructure. As of July 28, 2006, 275 U.S. cities have agreed to pursue the Kyoto Accord, despite the U.S. federal government's lack of insight about global warming (City of Seattle 2006a).

Another trend over the past several years, related to ecological cities and the recognition that cities do not exist by themselves but are part of a region, is an interest in *progressive regionalism*, a phrase that some planners are using to describe a revived interest in functional regions. In the past, functional regions have focused on the economic, much to the expense of the ecological and the social conditions. Progressive regionalism considers the intersection of local and global forces that express themselves at the regional level. A progressive approach to planning means a historically based yet forward looking, critical standpoint shared by people and organizations dedicated to eradicating root causes of poverty, social injustice/inequity, and environmental degradation. Included is a search for alternate forms of governance and ways to enrich civil society. It is not at all opposed to economic growth, but it provides a perspective that contextualizes economic growth in overall regional and personal well-being.

When connecting insights from progressive regionalism with insights about ecological city building, one develops a perspective that economy can thrive with ecology. Such perspectives counter arguments brought about by critics of progressive planning. Such critics indicate that the (American) planning scene is dominated by a *disconnected planning process* across space reinforced by an entrenched antiregional, political mentality. Many critics suggest that progressive regionalism will not work because of place-based circumstances. This textbook enters that debate head on, providing widespread evidence of *on the ground* regional community communication, cooperation, coordination, and collaboration directed toward overall contemporary and future well-being. This textbook encourages a look to the future as much as it talks about the present with regard to GIS use. Addressing this issue head on is to recognize that across the United States, local government plans and capital improvement programs are frequently disconnected; that is, plan and improvement programming databases are not commonly consistent, because the developments behind them are directed by different decision contexts. Integrating GIS databases to support decision process integration is a key issue for success.

2.4 Integrated Perspectives on Planning, Programming, and Implementation Decision Support

There should be an administrative link between plans, programs, and implementations. After all, projects out on the ground are the substance of capital improvement programs and capital improvement programs in turn implement long-range plans. That linkage has not usually occurred, because of the complexity of the individual processes. Many people are involved in each process. Many tasks are involved. A plurality of values is commonly involved—not always consistently. Nonetheless, the importance of the linkage is being recognized (Younger and O'Neill 1998; Meyer and Miller 2001).

To achieve some consistency in planning, programming, and implementation decision contexts, major concepts, such as integrated resource management, (smart) growth management, and sustainable development, have been proposed to address social, economic, and environmental conditions in a coordinated manner. Concerns about such conditions typically emerge when we consider the internal and external effects of land use, transportation, and water resource activities, as well as many other activities.

Plans (hence, programs and implementation strategies) can change, but not usually overnight. Implementation of projects changes more often than programs and programs change more often than plans. To address the linkage among plans, programs, and implementation to track the changes, it is important to be *conceptually consistent* among the processes. To do that, the GIS databases should be consistent in some way. Unfortunately, the data categories stored in GIS databases developed and used to address planning, programming, and project processes commonly do not link to one another, because they are *conceptually different* and decision situation needs differ. Therefore, it is rather difficult to implement integrated resource management, growth management, and/or sustainable development management from a broad-based perspective.

Importantly, conceptually relating databases within and between planning-, programming-, and implementation-level investigation processes (and logically linking databases as part of implementation of databases) occurs in two ways: (1) across functional elements and (2) across functional processes. Let us take, for example, the importance of linking (and in some respect integrating) functional elements in a growth management planning context, as in the State of Washington. Like many other states whose local jurisdictions plan under growth management regulations, Washington State mandates that local jurisdictions develop comprehensive plans that comprise several elements (e.g., land use, transportation, critical [water] resources, economic development) that influence one another. The nature of the plan elements and the influence among elements is left up to local jurisdictions to determine, because social, economic, and environmental conditions differ from jurisdiction to jurisdiction, both at the same scale and across scales. Each plan element can be considered a functional plan. Any pair of functional plans is likely to have some level of influence between pair members, because, in the real world, for example, land use—commercial/residential development—influences transportation (traffic volume/congestion), and transportation improvements (infrastructure) in turn influence land use (more or less intensity of use). Linking functional plans (as plan elements) is one way to take an integrated perspective with regard to growth management for the current generation in a community. Unfortunately, the level of functional linkage is still very much in its infancy. For example, what is known about the relationship between land use and transportation, or land use and water resources, or transportation and water resources, is not yet codified in GIS databases and the software used to manipulate those databases.

Another way to relate one dataset to another to foster insight for growth management is to link planning-, programming-, and implementation-level assessment. Plans are developed to guide growth. Programming is conducted to finance the build-out of a plan. Implementation-level investigation examines the details of the build-out in regard to impacts. Thus, overall, such a linkage among plan-, program-, and implementation-level investigation promotes a sustainable development perspective. The projects in a functional plan (e.g., a transportation plan) developed by the planning unit within an organization are proposed from the perspective of citizens' needs for the long term. These projects might be proposed based on land use development (i.e., growth in sectors of population, employment, and/or residential real estate). The projects that appear on a transportation capital improvement program, developed by a different unit within an organization, are those that are fundable over the next 6 years. Although organizations know that the projects in capital improvement programs are the same (or at least very similar) to the projects conceived in the planning stage, unfortunately, the planning projects and the programming projects commonly appear in disconnected databases, because the work processes between the two units of the local jurisdiction are done at separate times, under different direction. It is not always this way, but for most of the United States it is. Even more surprisingly, for most local jurisdictions in the United States, the databases for other types of capital improvement projects and planning projects are even more disconnected. The reason is that transportation is a fairly important issue to most people in most jurisdictions, and it attracts considerable financial resources, whereas parks, social services, and other less visible improvements attract fewer funds.

An enlightened perspective on geographic information integration is needed to support integrated resource management to pursue growth management operationally and/or sustainable development concerns. Information integration in itself is a complex endeavor—only recently being described with substantive, methodological, and conceptual depth. No wonder much of the world, particularly at the local jurisdictional scale, has difficulty developing a systematic approach to maintaining, let alone improving, the quality of life for citizens. The state of information technology is poised for doing a better job than we have been doing.

To illuminate the complexity of the situations and to cast a systematic light on the dilemma of what to know and how to interpret decision support information, we pose a framework for integrated situation assessment that comprises five dimensions (Table 2.4). The framework relates functional themes (activities), community conditions, decision process, spatial scale, and temporal scale. We can elucidate the character of these dimensions in a limited number of categories that to many are familiar.

An integrated situation assessment can be undertaken by taking these dimensions in pairs, as depicted in Table 2.5. Investigating any two columns or any two rows within the table constitutes an integrated situation assessment. One can choose as an analytic interest any two rows and/or columns simultaneously. Such a choice implies that the third, fourth, and fifth dimensions are taken as "control" dimensions (i.e., that is they do not change in the particular situation assessment at hand).

Although more dimensions and more factors within each dimension could be added to Table 2.4 (hence, Table 2.5 by extension), those depicted are the major dimensions for GIS analysis. The five dimensions are used throughout the text to differentiate decision situations. Taking them pairwise makes them more understandable, but all are important parts of decision situations, as described with the conceptual framework for decision situation assessment presented in Chapter 3. However, chief among the dimensions (in Tables 2.4 and 2.5) for understanding the use of GIS is the *decision support process* (i.e., planning, improvement programming, and implementation decision settings) that motivates and organizes GIS use in particular ways. Chapter 3 emphasizes the GIS-oriented decision support workflow process.

TABLE 2.4. Dimensions for Integrating Planning, Programming, and Implementation

1. Functional activities—land use, transportation, and water resources
2. Community conditions—social, economic, and environmental
3. Decision process scales—planning, programming, and implementation
4. Spatial scale—regional, county, citywide, small area
5. Temporal scale—strategic, tactical management, operational

TABLE 2.5. A Framework for Integrated Situation Assessment Comprising Five Dimensions (Dimensions Contained in Table 2.4)

	Community Conditions		
	Social	Economic	Environmental
Functional activities			
Land use			
Transportation		(9 elements)	
Water resources			

	Community conditions		
	Social	Economic	Environmental
Decision process scale			
Planning			
Improvement programming		(9 elements)	
Implementation			

	Functional activities		
	Land use	Transportation	Water resources
Decision process scale			
Planning			
Improvement programming		(9 elements)	
Implementation			

	Functional activities		
	Land use	Transportation	Water resources
Spatial scales			
Regional			
County		(12 elements)	
Citywide			
Small area			

	Decision process scale		
	Planning	Improvement programming	Implementation
Spatial scales			
Regional			
County		(12 elements)	
Citywide			
Small area			

	Functional activities		
	Land use	Transportation	Water resources
Temporal scales			
Strategic planning (long term)			
Tactical management (medium term)		(9 elements)	
Logistical operations (short term)			

2.5 Summary

GIS methods can be put to use quite effectively in several perspectives about urban–regional decision support situations. Five techniques are part of the conventional regulatory approaches to decision support: (1) community plans, (2) subdivision regulations and plans, (3) capital improvement programs, (4) zoning ordinances, and (5) public participation.

In planning-focused decision situations, people consider broad issues as part of the logic of community planning. Decisions that enter into a planning vision are typically for the long term, say, on the order of 10–20 years, but short-term plans get created in a number of communities. Long-term plans commonly address large areas, on the order of a city, whereas short-term plans address subareas when details are needed to firm up specifics.

An improvement programming decision situation considers what needs to change in regard to land, transportation, and/or water resources (and, for that matter anything else in a community that is budgeted for improvement), and how much the improvement is estimated to cost. As such, financing is an important part of the decision situation, in addition to all of the other factors considered in the plan. Consequently, an improvement program is a collection of projects, with each project having associated benefits and costs, but the benefits and costs are not necessarily expressed in monetary terms.

A project implementation decision situation is one in which a detailed economic and/or social, and/or environmental assessment is performed. At this level of decision situation, individuals (planners or consultants, etc.) who are responsible for the detailed analysis only have to consider the impacts related to a specific project. Given that the focus is narrow (in space and time), much more energy can be spent examining the details of how a particular project might impact a community. The project analysis might actually include several alternatives for a project. However, little, if any, cumulative impacts can be examined in this situation, because a single project is commonly the focus. Information from other projects is usually not considered—the law does not commonly require it.

Public groups serve to provide input, feedback, and recommendations about what governing bodies should do to foster collective choice in a democracy. The primary rationale for enhanced stakeholder participation in public land planning is based on the democratic maxim that those affected by a decision should participate directly in the decision-making process (Smith 1982; Parenteau 1988). Some GIS researchers have begun using participation frameworks to better understand how GIS can and should be used to foster public participation (e.g., the participation spectrum outlined by the International Association of Public Participation [2005], which involves moving from inform to consult, to involve, to collaborate, and to empower). These levels allow people to contribute more insights to growth management activities as we move from inform through midlevels to empower. The lower levels are more common than the higher levels.

The difference among conventional, growth management, and sustainability management is the degree to which functional themes, such as land, transportation and water resources, are considered separately or together. Conventional approaches consider them separately, whereas growth management approaches and sustainability management will consider them as interrelated phenomena. Furthermore, an administrative link between plans, programs, and implementations fosters sustainability management. After all, projects on the ground are the substance of capital improvement programs, and capital improvement programs in turn implement long-range plans. That linkage has not usually occurred because of the complexity of the individual processes. However, geographic information technology is continuing to mature to support an integrative perspective among the decision support situations. With that integrative perspective, governing bodies will be in a better position to address social, economic, and environmental conditions in a coordinated manner. Concerns about such conditions typically emerge when we consider the internal and external effects of land use, transportation, and water resource activities, as well as many other activities.

The five major dimensions for integrating planning, programming, and implementation decision situations include (1) functional activities, (2) community conditions, (3) decision process scales, (4) a spatial scale, and (5) a temporal scale. The most important of these dimensions for integrating planning-, programming-, and implementation-level investigation processes (and logically linking databases as part of implementation of databases) occurs in two ways: (1) across functional elements (activities), and (2) across functional processes. Due to the absence of dynamic process capabilities in GIS, the level of functional linkage is still very much in its infancy. As such, there is a considerable potential for improvement of geospatial information system technology that subsequently will foster improvements in conventional, growth management, and sustainability management decision support processes.

2.6 Review Questions

1. How do local governments regulate growth within communities through conventional approaches?

2. Differentiate between planning and improvement programming decision support situations in terms of the way projects are treated.

3. Differentiate between improvement programming and implementation decision support situations in terms of the way projects are treated.

4. In what way does a public participation framework help us understand the difference between weak and strong participatory democracy?

5. What are growth management approaches; and what is the difference between a top-down and bottom-up approach to growth management?

6. List and describe several GIS applications for growth management.

7. How can we compare and contrast growth management and sustainability management in terms of competing objectives and generational equity?

8. How might a decision process link among planning, programming, and implementation support sustainability management? Consider the competing objectives and generational equity in sustainability management.

9. What five dimensions might we use for integrating decision support situations?

10. Which of the five dimensions for decision situations provides the most leverage for integrating situations? Why?

CHAPTER 3

GIS Decision Support Methods and Workflow

GIS is known for its ability to integrate various types of data to address complex decision problems. When we undertake a GIS project, we make use of one or more methods. A *method* is a sequence of steps in which data are processed, commonly by a GIS analyst, to carry out *GIS workflow* to address a decision problem using various software components. Each method might contain one or more techniques for collecting, storing, preparing, analyzing, and displaying data. A technique is the basis for performing an operation on data. Some people use the term *capabilities*, and some software systems use the term *tool*. Rather than worry about the exact terms, we should sense what general steps unpack into more detail steps for performing operations. We pull these ideas together into a GIS (decision support) project.

In the context of this book, and particularly this chapter, we must be aware of a difference between a *GIS project* and a *community improvement project*. For example, a wastewater treatment facility is a community improvement project, but the process of siting such a project, as presented in the case study, provides a step-by-step method for undertaking a GIS project. The Green County wastewater facility project, which we use as a case study throughout this book, is related to but different than a GIS project: The difference is in the meaning of the term *project*. The wastewater treatment facility is an infrastructure project as part of a capital improvement program. The term *project* has a special significance when it comes to improving communities, particularly to planners, public works directors, council people, and the stakeholding public for that matter. The GIS project is the data processing activity (i.e., analysis of information to identify facility location). In fact, the term *project* is used so much that it will likely always require one or more adjectives to keep the idea straight, particularly in the context of planning, improvement programming, and project implementation decision situations. Thus, to be clear, here we define a GIS *project method* as workflow intentionally structured to carry out data processing activities. GIS projects make use of fundamental capabilities during a workflow. As such, we first describe the basic capabilities, then turn to workflow.

3.1 Overview of GIS Capabilities

Now that we have a way to differentiate decision situations, as provided in Chapter 2, we need to understand the types of GIS software capabilities we can use to perform decision support activities. Some people call such capabilities *functions* of a GIS. GIS software capabilities are subcategories of all the components (data, software, hardware, etc.) we must assemble when addressing a decision problem, as mentioned in Chapter 1. We use software capabilities to work with data, transforming data into information. As such, GIS software capabilities perform GIS-based transformations, whose purpose is to facilitate a transition from data to information by processing the elements of a database (as constituents of a map) and deriving their characteristics based on spatial properties of distance, direction, pattern, set membership, and nonspatial properties (attribute values). The elements of a database, called *features*, represent everyday world phenomena, such as rivers, lakes, streets, houses, utility lines, and so forth, or technical conceptualizations of everyday world phenomena, such as river reach, lake benthic layer, street centerline, house footprint, and so forth. In this book we confine ourselves to phenomena or their conceptualizations from the domains of water, transportation, and land resources. We discuss particular types of databases in Chapter 5.

In ArcGIS, distributed by Environmental Systems Research Institute of Redlands, California, the user has access to GIS capabilities through wizard-like *tools* (i.e., software implementations of specific functions). The tools are grouped into *toolsets*, which in turn comprise *toolboxes*. We cannot hope to provide an exhaustive list of tools herein, because the product brochures of most software vendors change from year to year. Instead, we offer a list of general categories of the most typical GIS capabilities (see Table 3.1). We can see that some toolboxes contain tools that operate on various aspects of the decision problem, content, structure, and process, but no tools actually operate on context as described in this section. Counting the number of entries (X) for each toolbox across the table columns reveals that data management tools and the map visualization tools are the most versatile in terms of addressing the information needs. No wonder many people view GIS as a map-based software technology. There are other functional categories as well. The geocoding tools, including linear referencing types of tools, establish coordinates for databases—a very important capability in GIS. The Feature Analysis Tools are likely to address information needs in situations characterized by spatial relationships among the discrete objects (e.g., presence–absence based on proximity, overlap, adjacency of boundaries, frequency of presence based on distance). The Grid Analysis Tools are likely to address the information needs in situations whose processes and relationships pertain to phenomena with fuzzy boundaries (e.g., soils, vegetation, terrain, pollution plume, noise) and continuous fields. Network Analysis Tools are likely to address information needs in situations whose processes and relation-

TABLE 3.1. A Framework for Understanding GIS Capabilities

	Content		Structure			Process	
	Features	Cells	Spatial	Logical	Temporal	Spatial	Temporal
Geocoding tools	×		×				
Data management tools	×	×	×	×	×	×	×
Map visualization tools	×	×	×	×	×	×	×
Feature analysis tools	×		×				
Grid analysis tools		×	×	×	×	×	×
Network analysis tools	×		×	×	×	×	

ships occur along edges (links) connecting to each other at nodes, such as movement along a transportation network, or flow through a drainage network.

In subsequent subsections we discuss each of six tool/transformation categories and their relevance to planning, improvement programming and project implementation decision situations. We describe the tools in the order in which a GIS analyst might encounter them by general category.

3.1.1 Geocoding Tools

Geocoding, also called *georeferencing*, is the process of assigning a location, usually in the form of coordinate data values, to a feature. Generally, anytime a GIS analyst assigns a coordinate to a feature, the analyst is geocoding. However, some systems designers like to reserve the term *geocoding* for assigning street addresses to coordinates. This latter usage, also referred to as reverse geocoding, is narrow, but it is nonetheless still correct. Technically, the assignment of coordinates to an address is accomplished by comparing the address description to the description in the reference database of addresses, and interpolating the address location, if needed. Addresses have many different formats, ranging from the common address of a house number followed by a street name and supplemental information to other location descriptions, such as postal zone or census tract. From the geocoding perspective, an address includes any type of information that distinguishes a place. The geocoding process involves potentially many tasks, such as the creation, maintenance, and deletion of address locators. The *address locator* defines the technique to be used by the geocoding process in matching addresses against the reference database. The geocoding toolbox contains tools to assist in accomplishing these tasks, tools that are relevant to planning scenarios, improvement programming, and project implementation decisions in which only address descriptions, but no coordinates, exist for entities comprising a plan or a specific project.

Linear referencing is a measuring system for linear features, such as river mile and route mileposts. These tools commonly are found in their own ArcGIS toolbox, but they belong to the same general class of functions as geocoding tools. A linear reference is a geocode along a linear feature (Nyerges 1990). The essence of linear referencing is to record location data by using relative positions along existing linear features. This is accomplished by using linear measures (e.g., kilometer or mile) from the starting point of a route. A route is a linear feature that has a unique identifier and measurement system stored with it. In graphic terms, a route represents a collection of line segments in one or more parts. Linear referencing allows multiple sets of attributes to be associated with existing linear features, such as speed limits, accidents, and pavement conditions with highway routes. These attributes can be displayed, queried, and analyzed with the tools of the linear referencing toolbox. Linear referencing tools are relevant to planning scenarios, improvement programs, and individual projects in the decision situations involving linear features (e.g., streams, roads, street centerlines) in which the feature attribute data are measured in a linear measuring system and the information about spatial relationships among the features is needed (e.g., the frequency of accidents between mileposts $x1$ and $x2$ along a divided lane road).

3.1.2 Data Management Tools

Data management tools offer capabilities to perform tasks that range from managing basic structures, such as fields and workspaces, to projecting and reprojecting coordinates of features/raster cells comprising a geodataset, to more complex tasks related to topology and versioning. The tools can be organized around toolsets, which are data management function–specific. For example, the fields toolset contains tools that make changes to the fields (attributes) in the tables of a *feature class* (that is, a collection of features with the same type of geometry and the same attributes sharing one georeferencing system); the joins toolset contains tools that add and remove a table join; the relationship classes toolset contains tools that create associations between feature classes, and between feature classes and tables; and the raster toolset contains tools that create and manipulate raster datasets. The relevance of data management tools for planning scenarios, improvement programs, and individual project decision situations is high, because almost any computation of plan and project impacts in GIS is bound to involve computations on entity attribute data.

3.1.3 Map Visualization Tools

Map visualization tools are the capabilities we use to compose displays. Although some GIS (e.g., ArcGIS) have a separate software module for performing map visualization, that packaging of tool capabilities is really a result of a product marketing strategy rather than overall functional similarity.

We can add, delete, and change data layers on a map. We can change the symbolization on a map. We can select and highlight features on a map by either location or attribute. Interestingly, we could not perform these activities if we did not have a data management capability working for us. This link between data management and map visualization is one of the fundamental advantages of data work in a GIS.

With map visualization tools we can pan, zoom, identify various features within a data layer, and measure distances between features. The data management tools support map visualization tools, and the map visualization tools support the data analysis tools. The capabilities are integrated among general functional categories, as mentioned in Chapter 1.

3.1.4 Feature Analysis Tools

Feature analysis tools offer processing for *feature* (vector data model) spatial relationships. Commonly, these contain toolsets, such as extract, overlay, proximity, and statistics.

Extract toolset contains tools employing attribute and/or spatial queries to extract features and their attributes. Attribute queries utilize relational (e.g., >, <, =) and/or Boolean (AND, OR, XOR) operators, whereas spatial queries utilize spatial relationships of distance, containment, overlap, and intersection to extract features.

The *overlay toolset* contains tools to combine, erase, modify, or update spatial features. All of these tools involve transforming two or more existing sets of features into a new, single set of features exposing spatial relationships between the input features.

The *proximity toolset* contains tools to determine the proximity of features within a feature set. These tools can identify features that are closest to one another and calculate the distances around and between them. A buffering tool that creates an exclusionary or inclusionary zone around a feature is a distance-based tool.

The *statistics toolset* contains tools for computing descriptive statistics on attribute data. The statistics include frequency count of each unique attribute value, mean, minimum, maximum, and standard deviation.

The tools of feature analysis are relevant for plan scenarios requiring the computation of spatial impacts and for a cartographic representation of real-world entities requiring vector features and spatial impacts that are the consequence of spatial relationships involving distance, containment, overlap, and adjacency. These tools are also relevant for improvement programming and project implementation decisions in which a project evaluation requires the computation of spatial impacts for individual projects.

3.1.5 Grid Analysis Tools

Grid analysis tools commonly have capabilities to analyze continuous surfaces represented by grid cell (raster) layers. The spatial analysis and data management transformations of continuous surfaces are packaged into toolsets. *Map algebra* is the language of spatial analysis for continuous surfaces, offering a set of functions for individual raster cells, cell neighborhoods, cell regions, and an entire raster layer. Some of these functions and operators (arithmetic, Boolean, and relational) were used to develop specialized analysis tools available in the groundwater, hydrology, interpolation, and surface toolsets. The tools commonly available for grid spatial analysis are relevant for planning scenarios, improvement programs, and individual project in decision situations in which the calculation of plan and project impacts requires continuous (raster) representation of spatial relationships and phenomena. Such relationships and phenomena include, for example, the association of vegetation types with elevation zones and soil types, or the gravitational movement of water over terrain surface.

3.1.6 Network Analysis Tools

Network analysis tools support GIS analysts who perform tracing along networks. There are many different kinds of networks based on the feature behavior. Electric networks are not the same as natural gas networks, which are not the same as highway networks or sewer networks; in fact, storm sewer networks are very different than water supply networks. Valves, transformers, and intersections are all very different kinds of junctions in networks. Analyzing flow with a grid analysis is different from analyzing flow with a network analysis tool, because the structure, given that the former is based on a raster data model and the latter is based on the vector data model (with topology).

When to apply certain tools from the toolsets is a matter of GIS *project workflow*, that is, how we should sequence the application of the GIS functionality. We turn to GIS project workflow next.

3.2 Workflow in GIS Projects

In this section about GIS project workflow we present the Green County wastewater treatment plant siting problem at three levels of workflow detail. First, we look at a simple GIS workflow, as presented and exemplified in the Environmental Systems Research Institute (ESRI; 2002) workbook—*Getting Started with ArcGIS*. Even a simple approach to workflow provides the foundation for GIS project learning, but it lacks insight about certain kinds of analysis. Because this book is written for an intermediate-level audience, we present in the next section a more nuanced GIS workflow. The more sophisticated one has been developed and used by Carl Steinitz (1990; Steinitz et al. 2003) over the last 20 years or so to address more fully geospatial problems. The simplified and the nuanced are synthesized into a third workflow that we believe is appropriate and useful for intermediate-level students. We synthesize these two workflows into an overall scheme that allows readers determine the effectiveness of workflow processes in generating information.

We presented in Chapter 1 two general, GIS-based workflow methods for integrating information to address growth management and sustainability management issues. One method is based on *data integration* across functional elements (i.e., integrate land use, transportation, and water resource themes for growth management). The other method is based on *process integration* (in addition to data integration) across planning–programming–implementation processes within one functional element for sustainability management. Undertaking information integration using a GIS-based workflow can improve the way resources are managed in public and/or private realms. Indeed, this book presents several examples of GIS-based workflow methods, many of which are fairly sophisticated. We come back to functional integration in Chapter 8, and to process integration in Chapter 9. Although it would be great simply to jump into the more advanced workflow methods, we first need to "walk" with GIS before we can "run." Whether we undertake data function–integrated GIS projects or process-integrated GIS projects, we first need to know more about GIS-based project workflow.

3.2.1 Basic Workflow for a GIS Project

A simplified workflow is based on an approach presented in the ESRI workbook—*Getting Started with ArcGIS* (2002). This *simplified workflow method* assumes that a decision problem can be solved by a single pass through a workflow, with limited testing of assumptions in each of four phases: (1) identify project objectives, (2) develop the database, (3) perform analysis, and (4) report the results. It is useful and instructive because of the simplifying assumptions regarding water flow process and impacts, for example, land use, transportation, and/or water resource movement, over space and time. The phases of project workflow are described in the following subsections.

3.2.1.1 Identify Project Objectives

We can ask a series of questions to articulate various objectives. What is the problem to be addressed? Who is the intended audience? Is the information for elected officials, technical specialists, and/or stakeholder groups? Will the data be used again? What final products are expected? Answers to these questions provide the guidance and information necessary to identify project objectives.

For example, the challenge for the Green County wastewater project is to find the most suitable location within the Green County community for a second wastewater treatment plant. The community has outgrown the current plant's treatment capacity, making a second facility necessary. Finding most suitable, less suitable, and unsuitable locations depends on the objectives for what constitutes a suitable parcel. Information provided to and discussed by the Green County Council suggests that a variety of site selection criteria need be considered to identify suitable sites. A target was set for each criterion (see left column of Table 3.2). Established criteria thresholds were based on discussions with the County Department of Public Works and the Department of Natural Resources. Knowledge about land uses and wastewater treatment facilities forms the basis of the problem definition in the Green County project and the basis for database design of the wastewater treatment plan in the GIS project. The middle column of Table 3.2 presents data most likely associated with the objectives of the wastewater treatment plant plan. The right column in Table 3.2 presents the data description agreed upon by the council, which is the basis for the data name in the middle column.

Studies of environmental controversies show that different stakeholder groups "value" different aspects of environments. As such, we might expect different criteria to arise, based on what is valued in any particular urban–regional community.

TABLE 3.2. Criteria for Siting a Wastewater Recycling Facility

Criteria data threshold for suitability	Data name	Description
Parcel within 1,500 feet of sewer lines	sewer	Areas near Green County (METRO) sewer lines
Parcel farther than 26,000 feet of treatment facilities	plant	Areas near existing wastewater treatment facilities
Parcel greater than 4 acres (174,240 square feet)	parcels	Green County parcels
Parcel farther than 500 feet from area susceptible to groundwater contamination	asgwc	Areas less susceptible to groundwater contamination
Parcel within the urban growth boundary	urban_growth	Urban growth boundary
Parcel within distance greater than 500 feet from school	schsite	Areas with less influence on Green County schools
Parcel within distance greater than 500 feet from park	parks	Areas with less influence on Green County parks and recreation areas
Parcel within distance greater than 500 feet from Green County critical slopes	slide	Areas less susceptible to landslide

3.2.1.2 Develop the Project Database

In a simplified approach to GIS workflow, the next step would be developing the database. Within this phase are three activities: (1) assembling data, (2) preparing data for analysis, and (3) organizing and storing the data for potential future use.

First, we assemble the data based on the existing database design. Second, we prepare existing data for use in the final analysis. These two steps comprise several substeps. Assembling data involves knowing the availability of suitable data from any source, including the organization from which the data can be obtained and the data model format for the particular features of interest, as obtained from various sources (see Table 3.3). We discuss data models in Chapter 5, but for our purpose here we call this a *format*, that is, the form of the data coming from an organization. Finding an organization source does not necessarily mean that we can acquire that data, but we try our best based on available resources (including money, time, and charm).

Data assembly can take the form of acquisition from primary and/or secondary sources. *Primary data acquisition* involves original measurements in the field, for example, taking global positioning system (GPS) survey points. We might have to partner with the source organization to help pay for the primary acquisition of the data, because it might not be readily available in digital form. However, most GIS work uses considerable secondary sources, because primary data acquisition is rather expensive. *Secondary data acquisition* sources are preexisting datasets. Although preexisting data might be available, this does not mean they are freely available. Some organization has had

TABLE 3.3. Data Availability and Formatting

Data name	Organization data source	Data model format
sewer	County Public Works	Shapefile
plant	County Public Works	Shapefile
parcels	County Tax Assessor	Shapefile
asgwc	County Department of Natural Resources and Parks	Shapefile
parks	County Department of Natural Resources and Parks	Shapefile
streets	County Department of Transportation	Shapefile
schsite	County GIS Center	Shapefile
slide	County Department of Environmental Services	Shapefile
wtrbdy	King County GIS Center	Shapefile
urban_growth	County Department of Environmental Services	Shapefile

to expend resources to obtain the data in digital form at some point. Consequently, many organizations are in cost recovery mode, thus requiring some type of asset value in return. This might take the form of an actual monetary outlay, a bartered in-kind service, or some future value to be negotiated. Once we have identified sources and obtained permission to use, GIS analysts can download data from database servers in *data warehouses* (large stores of digital data). The data model organization is important, because it provides basic information content for each of the data elements within each layer, and how this layer might need to be processed as part of the data preparation step; as mentioned earlier, data models are discussed in Chapter 5.

In the second step we prepare data for analysis. Some data come to GIS analysts in a form that requires intermediary processing, because they are not in a form that can be readily incorporated into a data analysis. For example, the coordinate systems of a particular dataset might not match what we would like to use; thus, we need to define a coordinate system for a dataset. In addition, to coordinate reference system compatibility, data format changes might be needed to facilitate GIS data analysis with certain data layers. More details about database design and manipulation are provided in Chapter 5.

Finally, we need to establish a procedure for maintaining the data in the data management system (e.g., ArcCatalog in the ArcGIS software). GIS analysts consult with a data administrator of the GIS unit responsible for data maintenance in their organizations to determine the best place to store the results. Because there are intermediate results in all GIS data processing, a significant amount of storage space might be needed.

3.2.1.3 Analyze the Data

The analysis involves performing various actions on the data to generate results that are appropriate to project objectives. Such actions may include geometric modeling (e.g., calculating distances, generating buffers, calculating area); coincidence modeling (e.g., overlaying data layers); adjacency modeling (e.g., path finding, nearest neighbor, and/or resource allocation).

In regard to the Green County project, there are two phases of analysis: preliminary analysis and final analysis. The preliminary analysis pulls together the data to reduce the dataset size. For the most part, this analysis involves clipping, that is, selecting a portion of the area and retaining that data as a final dataset (see Table 3.4).

The workflow for preliminary analysis is shown in Figure 3.1 (p. 48). The preliminary analysis makes use of the urban growth boundary (UGB) and matches that boundary against all the data. All features outside of the UGB are removed from consideration. The result for the parcels data is depicted in the map in Figure 3.2 (p. 49).

The final phase analysis focuses on the possible parcel set, in consideration of the large water users for a service area, and establishes a final dataset that can be considered a wastewater facility siting plan. Workflow steps included in the final analysis are depicted in Figure 3.3 (p. 50).

The resulting parcels for the final analysis are depicted in the map in Figure 3.4 (p. 51). The set of possible parcels is shown in light gray, and the most suitable parcels are shown in dark gray.

3.2.1.4 Report the Results

The results phase involves documenting the procedures in the form of a report for all other phases. A major part of the report shows the results of the analyses. Sketches can be made of the layout to orient map elements' placement. In the example of the Green County project, a sample sketch layout is shown in Figure 3.5 (p. 51). Because of the detail of information that can appear on maps, various trade-offs are appropriate with regard to detail and final output size. Some organizations might have access to large-format printers, thus making it easier to portray results.

TABLE 3.4. Data Processing Needed for Preliminary Analysis

Data name	Format	Processing needed
sewer	Shapefile data model	Clip to UGB
plant	Shapefile data model	Clip to UGB
parcel	Shapefile data model	Select parcels larger than 4 acres and clip to UGB
asgwc	Shapefile data model	Clip to UGB
park	Shapefile data model	Clip to UGB
freeways	Shapefile data model	Clip to UGB
schsite	Shapefile data model	Clip to UGB
slide	Shapefile data model	Clip to UGB
urban_growth	Shapefile data model	None

Note. UGB, urban growth boundary.

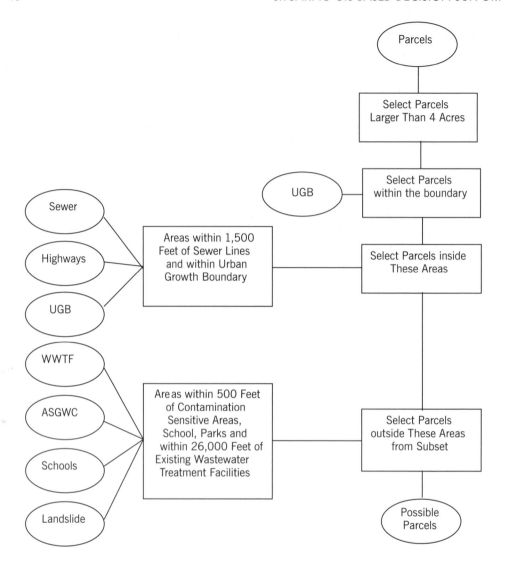

FIGURE 3.1. Preliminary analysis workflow for the Green County GIS project.

The previously discussed GIS project method is a start, but it would not be sufficient for many GIS development plans, programs, and project-level assessments, because we want to consider "real-world strength"—the ability of project results to withstand criticism. Thus, we need a more robust understanding of GIS project methodology to be complete.

3.2.2 Nuanced Workflow for a GIS Project

Below are phases in a landscape modeling workflow process formulated to address complex planning decision problems. The workflow process is comprehensive and robust enough to be applied in all three decision situations for planning, improvement programming, and project implementation. Emphases among the phases differ, of course, because the content and structure of a problem differ. The workflow process has been applied in practice in several GIS-related projects that address urban–regional landscape issues over the past decade or so, demonstrating a comprehensive approach to GIS work (Steinitz et al. 2003). We start each phase by considering/posing a number of issues/questions to focus that phase. We then apply the nuanced workflow to the Green County GIS project to enhance the reader's understanding of the breadth and depth of decision processes that underpin GIS work.

3.2.2.1 Representation Modeling

Several issues/questions that a GIS analyst should consider/ask in the representation modeling phase of a nuanced workflow include the following:

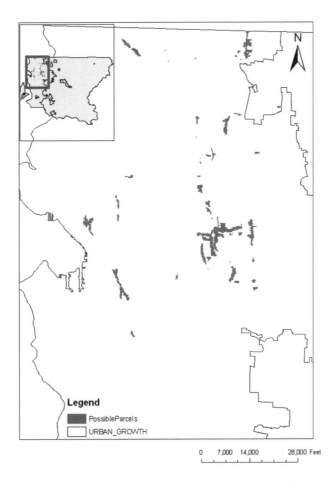

FIGURE 3.2. Parcels clipped to urban growth boundary, showing parcels for possible consideration in the site selection process.

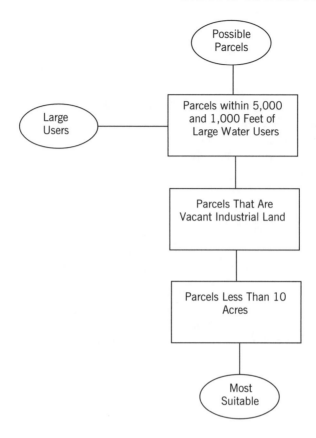

FIGURE 3.3. Final analysis workflow for the Green County GIS project, identifying suitable sites for a gray water treatment facility.

- How should the state of the urban–regional community, with regard to the particular issue at hand, be described in terms of a database design modeled as value trees or value hierarchies?
- What data categories are to be represented by measurements of attributes, space, and time?
- Whose concerns about these design questions should we consider? Should other groups be consulted to make sure we have incorporated all the relevant data into the representation model?

In a nuanced approach to GIS workflow we might expect that project objectives (e.g., in a simplified workflow) do not readily translate to a database design. In the Green County example, there appears to be some "magic" that moves the project from an idea about objectives to specific criteria used as a basis for database design. Thus, as steps toward demystifying that process, we present more details about database design issues in Chapter 5, where we discuss data models—foundation for data management systems— and database models—foundation for database design. Database models take advantage

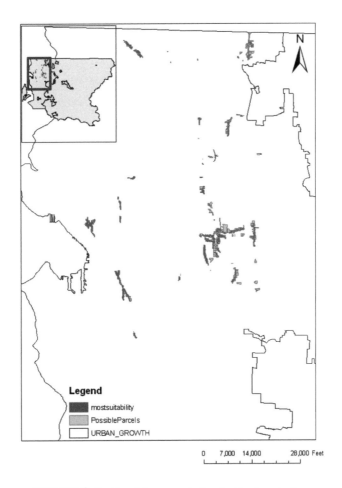

FIGURE 3.4. Resulting parcels for the final analysis.

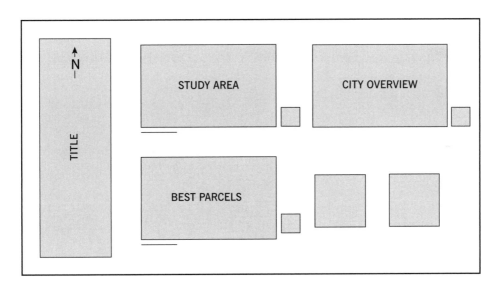

FIGURE 3.5. Layout sketch for poster map presentation.

of the foundational constructs in data models. Here, we provide only some highlights in regard to whose concerns appear in definitions of decision problems and, eventually, aspects of those concerns that get included in databases. These core concerns are the basis of the representation model through which the database model reaches full expression.

How did the Green County Council arrive at the set of criteria for addressing site selection listed in Tables 3.2–3.4? GIS analysts need to be aware of who values a wastewater treatment plant and why. The County Council seemed to arrive at a quite detailed set of criteria, without going through the preliminary step of negotiating priorities or values in the community. Complex decision problems are fraught with various interpretations of concerns about urban–regional communities. Stakeholder perspectives from diverse groups, even if these are groups within a single organization, tend to align with various concerns, often called *stakeholder interests*. These differences in interest are the basis of stakeholder groups.

Criteria (e.g., those in Table 3.3) develop out of concerns about the locational decision problems. For *any* data to appear in a database, someone had a concern that such data should be included. They took the time to design the database in a particular way, then collected (either by primary or secondary means) the data to populate the database. Unfortunately, many of these concerns are lost when they do not appear within the database. A *value structure* is an organizing framework for concerns (comprising values, goals, objectives, and criteria) that allows us to understand what we value in the world and how we can measure it. A value structure provides information about the *value of societal conditions* for interpreting the usefulness of geographic data, particularly an attribute associated at some time with a location.

Working in the context of transportation planning and decision making, Wachs and Schofer (1969) discovered that stakeholder interests for complex decision problems can be characterized in terms of value hierarchies of values, goals, objectives, and criteria. They described how certain abstract concerns need to be unpacked in more specific terms of values, goals, objectives, and criteria, if we are to make sense of complex (transportation) problems. They provided the following definitions and why it is important to relate values, goals, and objectives to criteria.

- *Value*—a basic desire and/or drive governing perception and behavior, whether inborn, instinctive, culture bound, and/or acquired by a person in growing and learning. Examples of values might be desire to survive, need to belong, need for order, need for security, and need for safety (Wachs and Schofer 1969 p. 135).

- *Goal*—idealized end state of the environment toward which people strive when they plan urban areas. Goals are generalized statements (but can be a clear, idealized end state) broadly related to physical environment values, but because they are abstractions, a test of fulfillment would clearly be difficult to apply. Examples of goals might be providing equal opportunity for all members of a community (related to belonging and security), or giving all members of a community access to transportation facilities that maximizes mobility. Furthermore, any given value might lead to conflicting goals. Deriving goals from values requires introspection and evaluation (Wachs and Schofer 1969 pp. 135–136).

- *Objective*—a specific statement that is the outgrowth of a goal and is truly attainable because of its reference to the physical world. An objective establishes a measured level toward which efforts are directed (e.g., the maximum allowable commute time from a particular distance to the central business district of a city incurred by a specific percentage of the population in a community; Wachs and Schofer 1969 p. 136). Here, again, a number of conflicting objectives might be associated with a particular goal. Note that this level of abstraction is defined in line with the planning literature rather than the decision science literature. In the decision science literature, the terms *goals* and *objectives* are reversed, which is why we have now appended the adjective *target* to precede the term *objective*.

- *Criterion*—the measurable characteristic associated with alternatives to be considered part of the solution to the decision problem. The level of measurement could be at any level (e.g., nominal, ordinal, interval, ratio; Wachs and Schofer 1969 p. 136). Criteria are the basis for qualifying and/or quantifying geospatial objective measurements. Being able to characterize geospatial data objects in relation to a value hierarchy can provide tremendous insight about how we interpret data in the database. As such, all data are *valued* in particular ways when it comes to interpreting the results of data analysis.

Independent of the work by Wachs and Schofer (1969), Edwards and vonWinterfeldt (1987) worked with a variety of stakeholder groups involved in oil leasing decision problems in the Santa Barbara Channel. They organized stakeholder interests into "value trees" to demonstrate similarities and differences among environmental, social, and economic objectives and criteria according to different stakeholder groups. These value trees organize objectives and criteria into meaningful chunks of information to compare what one stakeholder group might prefer compared to another. Edwards and vonWinterfeldt, together with Kenney, von Winterfeldt, and Eppel (1990), went on to use the value tree technique in several environmental problem-solving projects. The problem here is that the planning literature à la Wachs and Schofer (1969) and the decision sciences literature à la Edwards and von Winterfeldt (1987) provide opposite interpretations for similar words. The term *goal* in the planning literature means a broad-based statement of need to address a value. However, in the decision sciences literature, *goal* is interpreted as a specific measurement to describe an end state toward which we seek an outcome. The term *objective* also has different interpretations.

Building a database that recognizes peoples' *values* relative to the world is a step toward understanding what might be done to sustain and/or improve certain social, economic, and ecological conditions. There is a connection between community values and plans, and databases developed to provide a basis for data analysis to create those plans. In a pluralistic society, multiple values are common. It is important to understand how data might reflect certain "desired states of concern" about the social, economic, and/or ecological environment. We often measure what we value and value what we measure. Thus, because databases are stored measurements, we are led to analyze what we value and to value what we analyze. Consequently, we then map what we value and value what we map. As such, it is important to understand how databases, and the maps created from them, might perhaps reflect certain valued states of society and not others.

3.2.2.2 Process Modeling

Issues/questions that a GIS analyst should consider/ask in the process modeling phase of a nuanced workflow include the following:

- If a representation model forms a categorical content and structure foundation for a process model, then how might we examine relationships among land use, transportation, and environmental elements over time as a basis for articulating process?
- What are the relationships among the spatial–temporal elements, such as land use and transportation, that provide us with insight and a better understanding of urban–regional process?
- What land use, transportation, and or water resource processes do we need to consider?
- How do land use, transportation, and/or environmental transformation processes work?

In a process model it is important to track what influences what. Not only does sewage flow but also, once a plant is sited, there can be external plant effects on the surrounding area. It is important to understand (1) what is related to what, (2) what influences what, and 2) what causes what. These three issues in essence reflect strengths of relations among features and how they influence each other.

Urban–regional growth processes need be considered if we are to understand better how communities change. Porter (1997) characterizes growth in America's communities as being driven mostly by land use change. Land use change is driven mostly by economic conditions, but environmental constraints can and should be considered as well. Land use change is supported by access to transportation, because it is very difficult to get to places without transportation infrastructure in place. The land use and transportation relationship is fundamental in growth management. Land use and transportation are related to water, because land use activity "spills into water," and increasing land use intensity (from residential use to commercial use) generates vehicle trips on highways.

The siting of a Green County wastewater treatment plant is more a land use issue, but the operation of a wastewater plant is more a (waste)water issue. Nonetheless, the two are intertwined. Because land use activities have external effects (*externalities*), in the case of Green County wastewater treatment plant, the objectives articulated in Table 3.2 outline the relationships between the plant and the other described feature classes. The land parcel category, and the land parcel data within it, is the placeholder for the wastewater treatment plant, because the siting assignment is restricted to parcel location. For example, the plant should be located on land less than 365 meters in elevation. Commonly, it costs more money to pump water uphill than downhill. But we are not given any information about pumping (and other engineering information). The plant location (i.e., the land parcel on which it is located) should be outside the floodplain. If there were a spill, having the plant away from a low-lying area (in elevation) close to a river would prohibit the accidental spill from turning into a major disaster of sewage in the river. For example, at the end of March 2006, a major sewage pipe burst in Honolulu, Hawaii, spilling raw sewage into a canal. The canal was close to the shoreline and beach. Waikiki Beach was closed for 2 weeks, resulting in both the loss of millions of dollars in the tourist industry and many unhappy tourists.

A robust process model characterizes the *dynamics* of a problem, describing what features relate to other features across a landscape. It is meant to establish and to describe the fundamental relationships that lead certain features to *influence* other specific features for any particular category. A systems model description of an environment is an excellent way to implement a process model. A systems model accounts for key structural components (e.g., sources of sewage, treatment plant, and pipes routing the sewage), changes in characteristics of key components (e.g., sewage volume at different sources), and relationships among them (e.g., connectivity between sewage sources and the treatment plant). It can be used purely as a descriptive tool, such as the one in Figure 3.6 depicting structural relationships among the Green County feature classes.

Each arrow in Figure 3.6 can be made operational by refining the structural representations into formulas for computation. A systems model can be made operational in a systems dynamics modeling packages Stella or iThink (Ford 1999), but the spatial dimension is not included. A spatialized version called Spatial Modeling Environment was developed by Maxwell (2006) and modified to perform dynamic landscape simulation modeling. However, even if the descriptive version of the process model is never fully implemented in a systems modeling environment, the descriptive version still provides fundamental insight into what feature class influences/effects what other feature classes and can be used to inform the modeling process.

Taking the Green County example once again, the wastewater site selection problem should include the existing wastewater facility (if any) when choosing a location for a second facility. The process involved in the decision problem is the flow of wastewater through pipes that are underutilized, but facility site selection is also part of a growth management issue. The growth management issue includes information about vacant parcels that are likely to experience development, changing what is currently rural land to residential land. An expectation of urban–regional growth should be considered. Just how that growth occurs is based on urban growth process models.

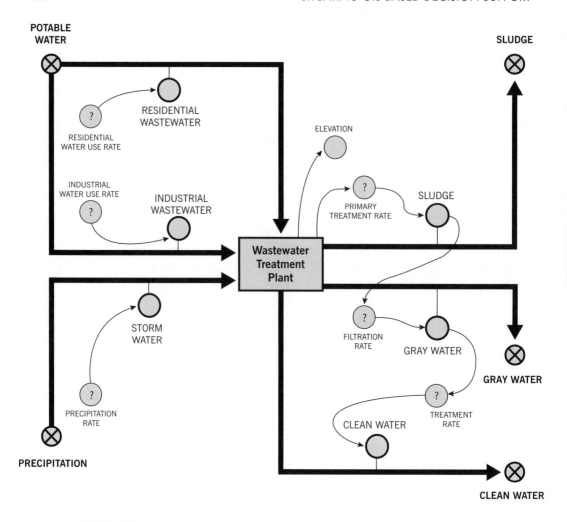

FIGURE 3.6. A model depicting water flow processes of the wastewater facility.

3.2.2.3 Scenario Modeling

Generally, several issues/questions should be considered/posed as part of the scenario phase of workflow:

- How does one judge whether the current state of the urban–regional environment is working well?
- What metrics of judgment (e.g., aesthetic beauty, habitat diversity, cost, nutrient flow, public health, public safety, and/or user satisfaction) are used to evaluate the nature of change?
- Which of these issues do we want to consider in a scenario? How many questions can people consider without getting lost within an information glut?

A process model forms a functional foundation for a scenario model (Figure 3.7). Scenario models are developed by *tweaking* assumptions about processes, because we can convert input into process. Given a different set of assumptions or inputs about how growth might occur, we can generate a variety of scenarios. Sometimes people refer to scenarios as the "worst case" or the "best case." Those references must be explicit about what "worst" and "best" mean. This recalls our understanding of the values, goals, objectives, and criteria that are part of representation modeling.

Each scenario is the result of a different set of assumptions about various values (e.g., safety vs. efficiency in the building of a wastewater facility). One should consider scenarios relevant to different stakeholder groups. There is likely no single scenario that everyone will prefer when entering a controversial discussion, such as where to locate

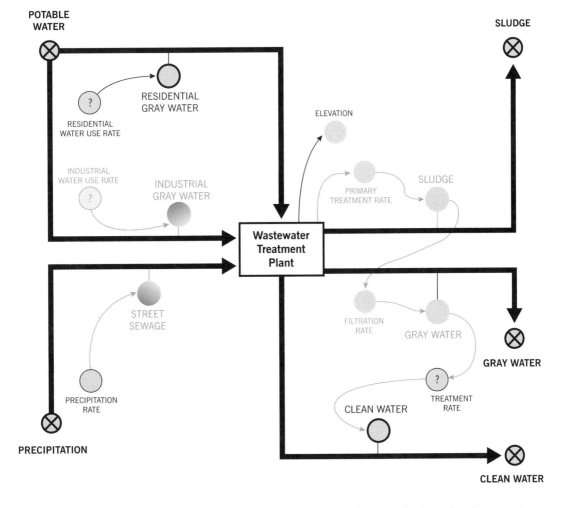

FIGURE 3.7. Important aspects of the process model graphic become the basis for the scenario model. Grayed-out elements are no longer an explicit part of the Green County decision problem.

a wastewater treatment plant. This is not to say a facility cannot be located eventually. However, it does mean that various scenario considerations are the basis of addressing spatial equity. Lober (1995) showed that each of five different fundamental assumptions about equity can result in a different locational choice for an incineration facility. The same is undoubtedly true for wastewater treatment, and any other controversial locational problem. Talen (1998) showed us a variety of maps that can be used to portray fairness in planning. We can use these ideas to create scenarios for the Green County wastewater project. If we change the criteria in terms of "distance to X," where X is any feature (e.g., a park), then we are generating a different scenario.

3.2.2.4 Change Modeling

Generally, there are several issues/questions to consider as part of a change model:

- By what actions might the current representation of the urban–regional landscape be altered, whether conserving or changing the landscape in regard to what, where, when, and so forth?
- At least two important types of change should be considered. The first is how the landscape might be changed by current trends. Modeling trends leads to projection models as the basis of change.
- The second is how a community might be changed by implementing design action. This leads us to developing intervention models as the basis of change. Intervention is a proactive approach to change.
- Again, how many variables can we consider in these models before becoming overwhelmed?

Scenario models provide a foundation for change models. Starting with the scenario model, we compute a "before" situation, also called a *baseline situation*, for the change model. We then allow changes to occur, and compute an "after" situation. The difference between the two is the basis of the change model, and these changes are cast in terms of social, economic, environmental, and so forth, conditions (see Plate 3.1). In this sense, the change model simulates outcomes of "what if" questions that we may pose.

Readers often wonder about the difference between a process model and a change model. The difference is that a process model is continuous, whereas a change model focuses on particular characteristics taken from the scenario model and recomputed with different parameter values representing different "what if" assumptions, so that we can incrementally detect the change of those specific characteristics in the scenario model.

Consider the benefit of building a wastewater treatment facility, in that many more people should be provided wastewater service and/or enhanced service. Computing the character of the change in service would be the basis of the change model. Furthermore, in the context of the Green County project, what changes might occur in population, residential water consumption, industrial water consumption, and so forth, that form the basis of a change model? The Green County wastewater project workflow does not consider those changes at all.

3.2.2.5 Impact Modeling

The following issues/questions as related to an impact model phase of a nuanced workflow should be considered/posed:

- A change model forms the basis of "what content, structure, and process" impacts to consider.
- What predictable impacts (i.e., the outcomes of changes) might those changes influence and/or cause?
- What impacts are less predictable because changes and processes are not well understood?

Although impact models follow from change models, impact models rely on high-quality information output from all of the preceding models. Impacts due to urban–regional growth—whether land use impacts, transportation impacts, or water resource impacts—are challenging to estimate. The difficulty arises from what remains unknown about the details of land use, transportation, and/or water resource processes. Although considerable geospatial data exist, when it comes to modeling impacts, we never seem to have enough of the *appropriate* data.

In the context of the Green County wastewater treatment facility, not all of the social, economic, and environmental impacts that result from siting a facility are known. Some impacts of criteria suggest that parcels are to be included for consideration, whereas others suggest that parcels are to be excluded from consideration. The criteria that sug-, gest a positive impact are called *inclusionary criteria*. The criteria that suggest a negative impact are called *exclusionary criteria*. Inclusionary criteria are commonly the inverse specifications for exclusionary criteria. The distance to the river, created by using a distance buffer, is an inclusionary criterion. Figure 3.8 shows an area for parcels before consideration of the buffer, and the influence of parcels that are not to be considered (showing the exclusionary aspect).

The objectives in the problem definition for Green County simply did not consider many of the social and economic impacts. We assume that the current wastewater treatment technology can clean the water to an appropriate level, wherever the facility is sited. However, noise-, air-, water-, and odor-oriented environmental impacts, in addition to financial challenges for the community, plus social disruption and esthetics, are not considered. Mitigation of those impacts must be addressed when a particular site is selected. To identify specific impacts in advance we develop an impact model.

Because analysis of impacts includes inclusionary and exclusionary analyses, we get non-ranked sites, as collections of sites that satisfy criteria when we perform decision analysis. To obtain ranked sites we have to compute gradations of inclusionary and exclusionary criteria, thereby ranking the effect that each inclusion and exclu-

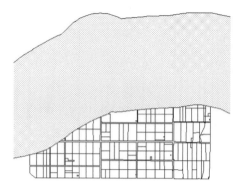

FIGURE 3.8. Land parcels outside the river buffer have been excluded from consideration, because they are too great a distance from the river for effluent discharge.

sion has on the final sites. Remember that information in this modeling process is a "feed forward" process, whereby every earlier step has an influence on the steps that follow.

3.2.2.6 Decision Modeling

For a long time, GIS has been touted as a decision support system (Cowen 1988). The nuanced workflow process makes this idea clearer because of inclusion of the decision modeling phase. One thing to remember is that data processing for all other models leads up to this model, but, in particular, an impact model forms the foundation for characterizing alternatives for a decision process. When we trade off one impact (computed and represented as an attribute) for another, we can set priorities for what we value. Chapter 7 provides more detail about these multicriteria considerations.

Generally, several issues/questions can be considered for a decision model:

- How is a decision to change or conserve the landscape to be made in regard to urban–regional impacts?
- How can a comparative evaluation based on a sensitivity of impact change be made among alternative courses of action?
- How are we to treat impacts in an equitable manner?

In the context of the Green County wastewater treatment facility project, we consider several criteria, each of which is concerned with social, economic, and environmental conditions. Distance to a park is a social consideration. Distance to residential areas is an economic (property value) consideration. Distance to river is an environmental consideration. Using a conventional GIS for implementing a decision model requires us to iterate through several analysis runs (Lowry, Miller, and Hepner 1995). This is the same as performing the Green County analysis several times under different scenario assump-

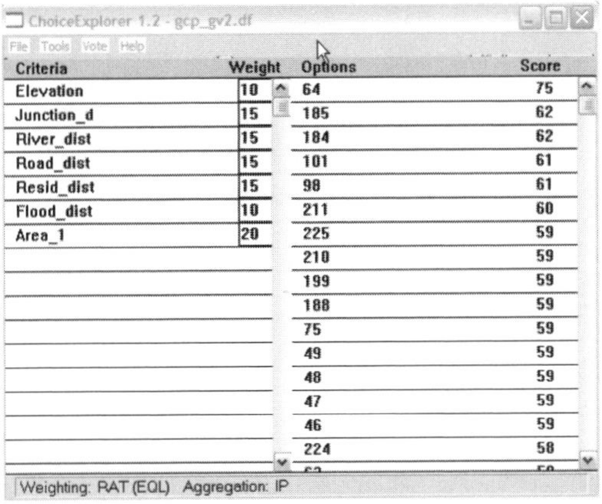

FIGURE 3.9. Weighting of objectives (site criteria), whereby area size is given the most weight (20 out of 100 points), and elevation and distance to floodplain the least (10), generates map in Figure 3.10, whereby site 64 is ranked the highest.

tions. It takes time to perform such an analysis, but is well worth the effort to see how robust a particular site selection outcome would be. If we use specialized software, like GeoChoicePerspectives (Jankowski and Nyerges 2001b), specifically designed for site selection interactivity, then we can rank and rerank the sites as appropriate to the nature of the "weights" of the criteria. In Figure 3.9 we show site rankings in a table format. In Figure 3.10 (p. 62) we show sites in a map format that corresponds to the table in Figure 3.9. These two displays are "coupled" in the software.

3.2.3 Synthesizing Basic and Nuanced Workflows

The landscape modeling framework just described adds several nuances to the simplified workflow, hence the name *nuanced workflow*. If a project workflow is more in line with the simplified version than with the nuanced version, then what will be the information outcomes of the simplified approach relative to the nuanced approach? When we compare the two workflows, we see there are similarities and differences in the phases (Table 3.5, first and second columns).

Although we draw a similarity between the project objectives phase in the simplified workflow and the representation modeling phase in the nuanced workflow, there are significant differences between the two. The project objectives phase of the simplified workflow process is meant to address problem concerns at the beginning of the process, as if people can readily do that. We identify the goals and objectives for the decision task at that point as if it were a "problem description." The representation modeling phase of the nuanced workflow involves elucidating the content (e.g., the physical characteristics of the community along with structural relationships among social, economic, and environmental characteristics that comprise the geographic decision problem). If the

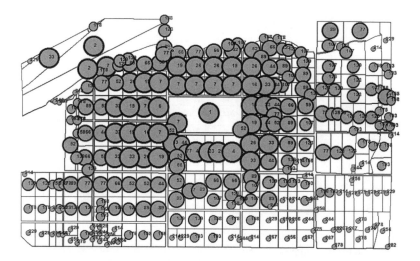

FIGURE 3.10. Site ranking based on the criteria weighting of Figure 3.9. Highest ranked site is in the upper center (rank of 1), and the next two sites are tied for second (rank of 2, upper left).

elements are not elucidated in the problem representation phase, then those elements cannot be considered as part of the process modeling phase.

Having compared the simplified and the nuanced workflows, we can synthesize an overall *recommended normative* workflow (third column of Table 3.5). This recommended workflow is the one we use below and in the following chapters. Adopting any particular workflow begs the question, "Does it really matter whether a GIS analyst performs all phases of nuanced workflow in a GIS project method?" Why or why not? If our choice is to perform only certain steps, is it due to a shortage of time? Is it due to insufficient understanding of the problem? Is it due to lack of data? Did our boss or instructor simply tell us not to do it? Is it just some or all of these issues? There could be myriad reasons for not doing a complete job. When this issue arises, an analyst should be aware that *every time we eliminate one of those six phases, we introduce additional informational uncertainty into the resulting informational product in a GIS project.* If we remove the process modeling phase and combine the change and impact modeling phases, then the nuanced workflow looks similar to the simplified workflow. Eliminating the process modeling phase means that we do not have insight into the details of process change that underlie scenario development. Eliminating change modeling leaves the impact models in a naive state: No information about change leaves the impact information uncertain.

Addressing uncertainty in a systematic manner throughout the modeling process can be accomplished through a three-pass approach (Steinitz et al. 2003). In each pass we make certain decisions about how much effort to invest in decision support. In the first pass, we consider steps 1–7 (in the third column of Table 3.5) in a way that allows us to become familiar with the scope and limits of the decision problem at hand. We recog-

TABLE 3.5. Comparing Basic and Nuanced Workflows to Derive a Synthesis

Simplified workflow	Nuanced workflow	Synthesized
1. Identify project objectives • Selection of criteria	1. Representation modeling; identify objectives in terms of all steps in workflow	1. Representation modeling 1.1. Problem description based on information needs expressed in terms of goals, objectives, targets (thresholds to reach), and criteria (data categories and measurements). Consider human resources for implementation in particular context. What is left unknown?
2. Create project database	1. Representation modeling, database development	1. Representation modeling 1.2. Database development—Specification and design of schema and implementation of database. Consider human resources for implementation in particular context. What is left unknown?
None	2. Process modeling	2. Process modeling—Identify critical relationships among features about how they interact. Consider availability of resources for implementation in particular context. What is left unknown?
3. Data analysis—single scenario based on inclusionary and exclusionary constraints	3. Scenario modeling	3. Scenario modeling—Select the characteristics relevant to various scenarios for a new waste treatment plant. Based on human resources available, consider the number of scenarios to compute. What is left unknown?
None	4. Change modeling	4. Change modeling—Compute the changes in the primary feature under consideration (e.g., number of people served by treatment plant at a given site). What is left unknown?
3. Data analysis—combined data layers	5. Impact modeling	5. Impact modeling—Given the change model, compute the external effects of siting a plant at the particular location. Site and situation impacts. What is left unknown?
3. Data analysis—single combination of impacts	6. Decision modeling	6. Decision modeling—Perform trade-off analysis using the impacts generated from the impact model. What is left unknown?
4. Report	None	7. Final report—Create the final report as a model of the information from all other steps. Use the interim reports from all other steps to synthesize a final report.

nize that each step enables and constrains the following step. A problem representation (e.g., what to consider in a GIS-based wastewater treatment facility project) suggests that if we do not represent people's concerns and/or the data available, then we cannot model process—or, for that matter, treat these concerns in all subsequent models. Subsequently, we cannot then compute change or model impacts, or use them to evaluate trade-offs to make a choice about where to locate a facility. When we scope out the final report, if we miss reporting some important information to people about the outcomes of the decision model, then that work will go unappreciated. A good report is the final chance to inform those who are interested in the information outcomes. What types of information products will be used? Maps, tables, graphs, and text all work together; these different conventions have been called *information structures*, because each has a different way of organizing and representing information (Nyerges 1991a). Empirical evidence about information structure use in decision settings shows that some people understand certain types of information structures more than others (Jankowski and Nyerges 2001a). We must make sure that we know our audience, and use a variety of maps, tables, and graphs to convey the information result.

This leads us to a second pass, which we apply backwards. In other words, whatever we would like to make a decision about in step 6, such as a facility location, requires that we have reasonable information to assess the impacts in step 5. Information in step 5 constrains what we can know in step 6 to make a choice about facility locations. Without impact information based on data, then, we are virtually guessing about the impacts. A good decision is based on evaluating appropriate information, such as that considered by stakeholder groups. In the same way, this means that we need data for assessing change in step 4 to compute reasonably and effectively compute the impact information in step 5. To assess change in step 4 we need to establish reasonably robust scenarios in step 3. Then we ask whether all the right processes have been considered. Thus, the scenario choices constrain the change modeling. Change modeling is constrained by how we model process about the problem. Are we modeling wastewater flows in a reasonably robust manner? To examine wastewater flow in step 2 we must have all the appropriate data in the problem representation of step 1.

The third pass involves implementation of each of the six steps. Thus, we should perform a robust database design and develop a robust database to have the appropriate data to undertake process, scenario, change, impact, and decision modeling steps. If we have not modeled the process sufficiently, then we should return to the problem representation phase to enhance the representation.

Understanding complex decision situations has been of significant interest to us for the past decade or more and is part of the motivation to write this book. A more effective understanding is one that addresses these additional nuances. Specific priorities are decided by each GIS analyst. One way to explore these priorities in detail is to use a framework for decision situation assessment that operates with the previously discussed workflows in mind—no matter how nuanced. The analyst decides how much of the context to take into consideration when performing a GIS project.

3.3 Summary

As mentioned in Chapter 2, and expanded upon in this chapter, GIS is touted as a decision support system. This chapter has presented many details of how and why GIS can assist with decision support. In this chapter we expanded on many of the details of decision situations presented in Chapter 2 by (1) describing many of the general capabilities of software and the ways these capabilities address decision problems, and (2) presenting different GIS workflows that make use of capabilities.

It is important to understand that the term *project* has at least two fundamental meanings when we talk about decision problems as the motivation in decision situations. One way to talk about projects is in terms of how communities want to plan, program, and implement land use, transportation, and/or water resource activities to restore, redevelop, or otherwise improve quality of life. Another way we have discussed projects is in terms of how we make use of GIS in a decision situation—the so-called GIS project. GIS analysts undertake GIS projects, working with data to address community concerns about community projects.

GIS is a special information technology because it integrates capabilities from three fundamental technologies—data management capabilities, spatial analysis capabilities, and map visualization capabilities—within a single system. Each technology, and its associated capabilities, alone is quite useful, but the integrated set is what makes GIS so special.

GIS capabilities are put to use in a workflow. Every time we use GIS, we establish a workflow. We introduced three workflows—simple, nuanced, and combined—to show that any two workflows, when compared, can generate at least one more workflow that perhaps might be more informed. The simple workflow (comprising four phases) is good enough to learn the basics of GIS. Another workflow that has shown considerable success in addressing complex landscape planning problems is one with six phases, introduced by Steinitz (1990). Each phase is a special modeling step. The third workflow is a synthesis of the other two, resulting in a seven-phase workflow. The main point is not simply to develop one more workflow, but to show that workflows can often take on emergent qualities to address special nuances of decision problems. We make use of the workflow characterization in varying detail throughout the rest of the book.

3.4 Review Questions

1. Why does it matter that you understand the difference between a GIS project and a capital improvement project when talking with people other than those familiar with your GIS project?

2. What is the relationship between the three perspectives of GIS embedded within the GIS definition in Chapter 1 and components of software presented in this chapter?

3. What are the major software capabilities of GIS?

4. Why use content, structure, and process to describe GIS tools?

5. How would you describe a basic GIS-based workflow approach?

6. What is the difference between a preliminary analysis and a final analysis?

7. How would you describe a nuanced, GIS-based workflow approach?

8. Describe the difference between the concepts of value, goal, objective, and criterion.

9. Which concept—value, goal, objective, or criterion—is most similar to an attribute in a GIS database? Why?

10. What is the difference between impact modeling and decision modeling?

CHAPTER 4

GIS-Based Decision Situation Assessment

This chapter presents two frameworks that can help decision analysts better under-stand the character of decision support. The first is a framework for differentiating sim-ple, difficult, complicated, and complex decision problems. Solving complex decision problems motivates analysts to use GIS, or some other information technology that has only some of the capabilities of GIS. In the second framework, many aspects of decision support are pulled together into a "big picture" of decision problems and workflow by introducing readers to what we call *decision situation assessment,* which is a process that allows GIS analysts to think through what is needed to address GIS decision support problems—not only from an information perspective but also from a broad-based orga-nizational perspective. We describe how a GIS decision analyst can perform decision situation assessment at four levels of detail—general, phase, phase–construct, phase–construct–aspect—and thereby uncover much of the complexity in planning, improve-ment programming, and implementation decision situations that were introduced in previous chapters. We hope that uncovering decision situation complexity fosters a more informed GIS-based decision support process, but, of course, that depends largely on the skills and expertise of the GIS decision analyst.

4.1 Characterizing Complex Decision Problems

Several researchers have described complex planning decision problems in various ways. The terms *structured, semistructured/ill-structured,* and *unstructured (decision) problems* have been used. They are great concepts but troublesome to make operational; thus, they are difficult to use. Many people can understand the following four terms: *simple,* *difficult, complicated,* and *complex.* They might agree that there are differences in these terms. However, here we can deepen that understanding through a framework about complex problems (Table 4.1).

**TABLE 4.1. Characterizing Decision Problems in Terms
of Systems Components**

| Type of decision problem | Four decision problem components in an open system | | | |
| | Three decision problem components in a closed system | | | |
	Content	Structure	Process	Context
Simple	×			
Difficult (semistructured)	×	×		
Complicated (ill structured)	×	×	×	
Complex (wicked)	×	×	×	×

We can make use of the problem framework by considering a decision problem about a wastewater facility location, for example, a water recycling facility site selection in Green County. One way to use the framework is to consider the GIS to be part of a decision support system. If we can enumerate all parts of a system, then we have a *closed* system. If we cannot enumerate all parts of a system, then we have an *open* system. It is much easier to solve closed-system problems, because there are no outside contingencies. Open-system problems always involve contingencies (e.g., politics and people's interpretations of problems). However, we can still solve open-system problems by enumerating sufficient parts of such a system to be comfortable with the nature of the solution. If we use the framework to address a wastewater problem dealing with wastewater flow, then we describe the component parts of a wastewater routing system, such as water sources, pipes, treatment facilities, outflows, and water sinks. The *content* of the problem deals with all the water-oriented things; but we could even include people (or at least important decision points). The *structure* is the relationship between those things, such as pipes connected to outflows. The *process* is how the water moves through the system. The *context* involves environmental or social factors that might influence water flow. Together these components constitute the wastewater flow problem.

Another way to use the framework is to embed the wastewater facility site selection in a question of land use, so that the land use problem becomes a decision problem. The two problems are intertwined and, in fact, are both part of the overall wastewater treatment improvement problem that a community may face. Understanding the land use problem (even if we do not take this to be a wastewater problem) is at the core of the wastewater facility site selection. The *content* of the decision problem becomes the land use change as a location problem. The *structure* of the decision problem becomes a matter of how to organize the decision components we use to guide the site selection. The decision *process* of the decision problem is the workflow we use to compute the necessary information to accomplish site selection. The *context* for the decision problem involves all the individuals and extenuating circumstances surrounding the solution to the decision problem. For example, some people might care more than others about getting the site selection *right*, such as the neighborhood groups who live closer to the facility sites and are more interested in the locational problem than are people living farther away.

Characterizing GIS projects in terms of content, structure, process, and context is rather important. Remember, there is a difference between a *GIS project* and a *community improvement project*. Be aware of the importance of adjectives, because there are many nuances for the term *project*. There are many ways to use the term, and those ways imply a variety of nuances for undertaking GIS work. Let us be a bit clearer about the differences behind decision problem *content* before we get to the topic of GIS workflow *process*.

In thinking about locating the wastewater facility site, we ask the question: What is the content of the Green County GIS project really about? It is, after all, a constructed GIS problem, but considering the coordinates being used, the data are from an actual place; thus, in many respects, these data are realistic. From the *content* category in Table 4.1 (which is the same as the *functional activities* dimension of situation analysis framework presented in Table 2.5), we want to know: What kind of decision problem in terms of substance are we facing? Is it a land use, transportation, and/or water resource problem? The Green County GIS project introduced in Chapter 3 involves a land use decision problem more than a water decision problem; consequently we suggest that students consider the GIS project to be a land use and water project to be more realistic. The Green County project has little in the way of water resource analysis, because wastewater flow is not considered at all, but it should, because flow significantly impacts the capacity of the system. Water situation analysis in the Green County decision problem is largely absent; land use change is the real *content* focus.

Using the dimensions of the situation framework presented in Chapter 2 and the process workflow outline of Chapter 3, we now ask: What kind of decision *process* is contained in the Green County project? That is, does the GIS project more closely follow a land use planning process, land use improvement programming process, and/or land use project implementation process? We ask each question in turn.

- Is the Green County GIS project performed in support of a planning process?
    - Yes, in the sense that the GIS project takes a broad-based, rather detailed examination of the siting process. We assume the city has conducted a wastewater needs analysis, including identification of potentially appropriate areas for siting a treatment plant, because this normally is done as part of a wastewater planning process.
    - A GIS analyst is directed to consider parcels only within a subset area of the entire city: where, we assume, the wastewater need is greatest, or will be in the future. However, we do not know this for sure.
- Is the Green County GIS project performed in support of improvement programming process?
    - No, because budgeting is not in question.
    - On the other hand, we can also say yes, because we are creating a prior-

ity list of sites that seem suitable and selecting among them. Improvement programming process not only includes budgets but it also furthers project prioritization. However, we are not considering the trade-offs among sites in a direct manner, although our analysis gives us some information whereby we could do that.

- Is the Green County GIS project in support of project implementation?
  - That the Green County project does not examine details of a particular site as much as it considers the differences among sites would indicate that it does not support the project implementation process.
  - If we were to consider the Green County project for capital improvement implementation, then we would consider the attributes of a particular chosen site, look into the details of that site (e.g., seismic stability, soil contamination, mitigation for construction noise, and other issues) as part of a capital improvement project implementation process.

Consequently, we can say that the Green County wastewater treatment plant site selection can be best characterized as a "land use problem that employs a small-area planning process." In some sense, it is a cross between (1) a planning process, because we are considering lots of information but none in considerable detail; (2) an improvement program process, because we are making a short-ranked list of sites, although without understanding trade-offs; and (3) an implementation process, because we are looking to understand the suitability of a particular site, but that suitability is not synonymous with whether capability (in a engineering sense) of building on the site is the right thing to do. However, site selection is mostly about data analysis in support of a *planning* process.

4.2 Decision Situation Assessment for a GIS Project

All four of the content, structure, process, and context components are important for characterizing complex decision situations, but those components do not help much with many *contingent aspects* of decision situations. Contingent aspects of a decision situation might or might not introduce complicating/complex issues into decision processes and outcomes, particularly within an information technology setting. Use of the term *contingent aspects* is another way to say that something is dependent on something else. We might ask: Is there any way systematically to understand such contingent aspects of a complex decision problem, when some of the concerns seem to be outside of the problem itself (e.g., organizational concerns, personnel concerns, or even information technology concerns)? Some people suggest that such concerns are really not part of a problem; but such concerns influence how we interpret and respond to core problem issues.

We perform decision situation assessments to understand better the circumstances surrounding the use of information, particularly when decision situations are complex

(Jankowski and Nyerges 2001a; Tuthill 2002; Miles 2004; Ramsey 2004). We have developed a decision situation assessment framework over the past several years based on a theory of group-based GIS use called enhanced adaptive structuration theory (Nyerges and Jankowski 1997; Jankowski and Nyerges 2001a). EAST was developed from a synthesis of adaptive structuration theory (DeSanctis and Poole 1994) with 14 other frameworks. AST contributed over one-half of the characteristics, which is why the resulting name became enhanced AST.

EAST has been applied in numerous contexts, such as public health, transportation, and habitat redevelopment decision support (Jankowski and Nyerges 2001a), hazardous waste cleanup decision support (Drew 2002), water resource planning decision support (Tuthill 2002; Ramsey 2004), and earthquake-induced landslide decision support (Miles 2004). A decision situation assessment can be performed at four levels of detail:

1. General decision situation assessment—Has GIS been useful/is GIS useful/ or can GIS be useful at all in describing the geospatial situation by convening (input), process, and outcome concerns associated with information use?

2. Decision situation assessment by phases—Has GIS been useful/is GIS useful/or can it be useful in a phase-to-phase description of convening (input), process, and outcome concerns associated with geospatial information use?

3. Decision situation assessment by phase–constructs—Has GIS been useful/is GIS useful/or can it be useful in describing all (or a selected set of) constructs within each (or a selected set of) phases associated with geospatial information use?

4. Decision situation assessment by phase–constructs–aspects—Has GIS been useful/is GIS useful/or can it be useful for describing all (or a selected set of) aspects within all (or selected set of) constructs, within each phase (or a selected set of phases) associated with using geospatial information?

Decision situation assessments differ in terms of the amount of work they require. The amount of insight gained differs according to the amount of work invested. We can generalize across the four levels, but every analyst might know more or less about a topic, so it is up to the analyst to decide how much effort to put into the assessment. Through these four levels we come to appreciate better that different audiences and purposes exist for the different details of an assessment. Audience, purpose, and level of human resource effort *really matter*; hence, we should be clear about them before we embark on an assessment. It is important to understand that these four levels can actually be customized to fit the needs of an analyst. The first level is most general and takes the least amount of effort. The second level expands on the first level. If we need more information about something, for example, about a particular phase of GIS work, then we go to the second level of detail. The third level expands on the second. If we want to know about the characteristics of a particular phase, then we use the phase and construct level. Finally, if we are still unsure, because many GIS projects can be quite complex, then we go to the fourth level and ask ourselves as well as others how to view the different aspects (there are 25 aspects) presented below for each phase of a project. Overall we can suggest to the reader that the effort needed and the outcome of any assessment performed, regardless of level, depends on what a GIS analyst knows about the GIS project, and whether the analyst needs to ask penetrating and/or insightful questions.

In the subsections below we expand on the usefulness of the decision assessment activity. By example, for each description of the four levels of assessment, we adopt the decision situation for the Green County GIS project and make reference to the various GIS workflow processes as appropriate. As we do this, we make reference to the uncertainty involved in knowing how to perform the GIS analysis. The uncertainty issue ties back into the different levels of detail in the workflow processes we presented earlier. The less we know about the work involved in the decision support process, the more questions we should be asking—whether this questioning be of oneself or of someone else to ensure that we are effectively addressing the decision support problems. Not knowing how to undertake GIS work constitutes a risk in the outcome of a GIS analyst's work. Sayer (1984) would have us be practically adequate about the risk—practically adequate to the situation. The risk of not knowing could result in a "lousy job for the day," such that no one is much assured that we can do it again tomorrow. It could be that our employment depends on our knowledge and on doing a good job. Ultimately, a life might be at stake—be it plant or animal—or very valuable resources might be at stake. The choices we make about needing to know more information depend on the circumstances. The questions involved in decision situation assessment can help us in our efforts to know something (since we can never know everything) about decision support situations.

4.2.1 Decision Situation Assessment Using a General Approach

A general assessment provides us with the "laugh test." That is, if GIS has been used or is being considered for use, there are certain core issues that, once addressed, should help to avoid the circumstance of someone "laughing out loud and saying, 'You used GIS to do what?'" Let us take as an example the Green County wastewater facility site selection project. Can the project be done without a GIS? Indeed, wastewater facilities for years have been sited without the use of GIS. However, in this day and age, would any organization *not* use GIS?

In a general approach to decision situation assessment, there are three main concerns to consider. First, what are the concerns about *convening* a decision situation? We could also ask about the input to the decision process. Second, what are the concerns about the *process* involved in a decision situation? What is the nature of the process? Do we really need to use a GIS to make a site selection? Why not just get a bunch of maps for the area, have a look at them, and decide? Third, what are the concerns about the *outcomes* of a decision situation? Is any site just as good as any other? Would we have to defend a decision outcome, for example, to the public or in court? The relevance of geographic information for each of these three main concerns is an important issue.

In regard to the concerns about convening a decision situation, we answer questions about the mandates, problem, people, and information technology involved. Is geospatial information at all relevant to any of these? For example, does a law mandate that policy be implemented in a certain way based on the population distribution, whether the population is people, fish, flora, and so forth? The National Environmental Policy Act (NEPA; 1970) constrains federal action to protect critical resource areas when federal money is used for planning, programming, or projects. The Washington State Environmental Policy Act (WA SEPA; 1970) constrains state action to protect critical resource areas when state money is used for planning, programming, or projects. Were any of these questions asked in the Green County GIS project? It is most likely that the council members knew of laws, which is how they determined what objectives to pursue.

Would siting a wastewater treatment facility motivate certain conflicting perspectives from various groups? Additionally, what kind of information and/or people might we need? What data are to be used? For that matter, what information technology do we have at hand to address the decision problem? Which of these questions might have been asked to get the Green County GIS project started?

In regard to concerns about decision *process* in a decision situation, is the process meant to be single phase or multiphase? Single-phase processes are often too short to warrant the use of information technology. However, if the single phase is very long and extended, then perhaps managing geospatial information in a GIS is needed to enhance organizational memory; that is, with turnover in staff, a geospatial database can reduce the knowledge loss of what is known about complex decision problems.

By example, is the Green County GIS project single phase or multiphase? If it is multiphase, how many phases comprise the project? How many people actually worked on the Green County project? Was only one person involved, such that every step involved was known by a single individual—even if most (if not all) steps are described in the workflow in this textbook?

In regard to concerns about outcomes of a decision situation, it is advantageous to depict geospatial information in map form, so that people can interact with a representation of the decision outcome. A map is a great conversation generator, whether the topic is controversial or not. Did the reader consider different map layouts for the Green County project depending on the audience considered? Was a map the only outcome of the GIS project?

The preceding concerns should not be considered different phases of a decision situation, although they commonly are when decision support is considered from a narrow context, as in most GIS textbooks. We really need to understand the decision situation in terms of preconditions of the process (i.e., our convening concerns), the process itself (i.e., the work we do), and postconditions of a process (i.e., the outcomes). As such, we were speaking about a single-phase decision situation. This is undoubtedly an oversimplification, because complex decision situations comprise multiple phases, such as the simplified, nuanced, and synthesized workflow phases described in Table 3.5. Therefore, we conclude that there is more to do to understand about the aspects of the Green

County GIS project. But even if we know that there is more to uncover about the Green County decision situation, we can understand that even a generalized decision situation assessment is better than none at all. So let us not stop here. Let us ask more insightful questions about what we could/should know about the Green County GIS project.

4.2.2 Decision Situation Assessment by Phases

A decision situation assessment conducted by phases makes use of the same activity we discussed earlier but now asks the same three questions for each of the phases rather than for the entire decision situation all at once. This is called *iteration*. We iterate for each phase by using the same three questions:

1. What are the concerns about convening this phase?
2. What are the concerns about the process in this phase?
3. What are the concerns about the outcomes of this phase?

For example, in the Green County GIS project we described a simplified workflow presented by the Environmental Systems Research Institute (ESRI) as comprising four phases: identify project objectives, create a database, analyze data, and report. For each of the phases we can pose the three questions mentioned above. The reader can see that the answers are undoubtedly different among the phases. Thus, asking these three basic questions for the entire decision situation as if it had only one phase could develop the same answers as a friend asking the reader "What do you do at work?" However, that answer would not be sufficient to address the details of the GIS project adequately to get the work done. Upon further examination, a reader might conclude that the answers developed for the general assessment actually do apply to the phase-based assessment. In fact, the questions "What are the concerns about convening this decision situation?" and "What are the concerns about convening the phase identified objectives?" can result in very similar answers. Furthermore, the questions "What are the concerns about the outcomes of this decision situation?" and "What are the concerns about the outcomes of this phase?" also could result in very similar answers; that is, the first and last questions of the assessment can generate similar answers. However, consider the number of intermediary questions. Comparing the numbers of questions between the different assessments, we have three questions in the general assessment, and three times the number of phases (four in the simple workflow) gives 12 questions in the phase assessment. Thus, the middle question of the general assessment is carrying the load of 10 questions in the phase assessment. We can conclude that the phase assessment will likely provide more insight.

Furthermore, consider the different number of phases between the simplified workflow (four phases) and the nuanced workflow (six phases), and the synthesized workflow (seven phases). We clearly showed a difference between the uncertainty of information in the simplified workflow and the nuanced workflow. Certain models (e.g., the process model and impact model) were very simple or were not even created in the simplified workflow. Thus, asking questions about what information is needed from each of these workflow phases is clearly to the analyst's advantage. Additionally, asking about the nature of the map output (e.g., the report phase that showed up in the simplified and synthesized workflow) is also very advantageous, assuming that an analyst does not know about all information that is desired by his or her audience.

Although we encourage a much more comprehensive approach to decision situation assessment in the phase approach than in the general approach, there are two important issues to address. First, do we have the time to do all this? Second, do we really have all the information needed to perform the work (i.e., is our outcome likely to be adequate to the need for information)? To answer the first issue we can say, only do the needed phase assessment. Who says we have to do every phase this way? We do have to perform the work. If we get a miscue here and there, we can still do it over again. So, we perform only the assessment really needed for the situation. To answer the second question, we can say that decision support is actually more complex than asking just three questions in an iterative manner. We now turn to an assessment by phase and construct to provide still further insight for the kinds of questions to ask—but only when necessary. It is necessary only when an analyst really feels uncomfortable about what he or she knows, or when he or she is really curious about how the information might be used, that is, when the analyst wants to know the slightly bigger and deeper picture about decision support.

4.2.3 Decision Situation Assessment by Constructs within a Phase

In a decision situation assessment by phase and construct we now must turn to the body of knowledge behind the first two assessment levels, as well as that behind this level and the next. We have been borrowing from a theory of GIS-based collaborative decision making called EAST, developed to explain how people work with geographic information in complex decision situations (Nyerges and Jankowski 1997). The core concepts in the theory are convening concerns, process concerns, and outcome concerns, each of which can be described in more detail using *constructs* (See Table 4.2) by phase. Do not be bothered by the term *construct* so much; it is just another word. However, do understand that convening concerns have at least three constructs: motivation from social-institutional mandates; people involved in the decision problem; and technology used to address the problem. Process concerns involve at least three constructs: initially adopting some aspect of the convening constructs (it is called *appropriation*); using analysis and deliberation in group process; and other things we might not expect (called *emergent concerns* about mandates, knowledge, technology, etc.). Outcome concerns involve at least two constructs: decision task outcomes and social outcomes.

For example, for each phase in the six-phase nuanced GIS workflow process, we can ask ourselves and/or others about each of the eight constructs in Figure 4.1 in the context of the Green County GIS project. If we asked all of these questions at this time, clearly you might take a snooze, if you have not already. In addition, that much description would take considerable page space in this text, so, as an example, we perform the

TABLE 4.2. Decision Situation Assessment: Representation Modeling Phase by Construct

Concerns and construct	Description of the concerns in problem representation
Convening concerns/constructs	
1. Motivation from social-institutional mandates	What laws, regulations, and directives do we have that tell us what to consider for the "decision problem (representation)" in Green County? The council members told us what objectives to use; thus, we assume that they know the law—or at least we hope they know what is lawful.
2. Group-participant knowledge	What knowledge do participants in the situation have about the problem, and how is it best tapped? Do we, as analysts, have the knowledge needed to undertake the database design for the Green County project? We are given the database design, so it is made easy. What if we were not given the database design? To whom do we turn for answers to our questions about what data to use? We should explore similar decision problems faced by other communities to see how they addressed the problem.
3. Participatory GIS technology	What database design capabilities will we need to address the wastewater treatment facility siting issue? Are there any tools to which we might or might not have access that would make the job easier? Let us assume that ArcGIS is a useful toolkit to use for the GIS project, so what tools might we use for database design? Can we wait to read Chapters 4 and 5 before we know a lot more about this?
Process concerns/constructs	
4. Appropriation of social-institutional mandates, knowledge, and technology	Should we appropriate all of the information we have discovered by considering the convening constructs? Or is there some priority information set to start this problem representation phase, then follow through with other information on an as-needed basis during the GIS project? Who says we need to have all data up front? Why bother?
5. Analytic and/or deliberative group process	What is the nature of the task activity within the problem representation phase? Do we actually know enough about database design to carry through on this task? What combination of technical steps and steps facilitating a discussion are needed to fulfill the information need for this phase?
6. Emergent influences from mandates, knowledge, and technology	What new insight might be gained from adopting institutional mandates, new knowledge, and/or new information technology to address the issue at hand in this phase the same as others might have addressed it? Did we learn anything while talking to each other about this phase that we did not know before we started the phase?
Outcome concerns/constructs	
7. Task outcomes	What are the expected information outcomes from this phase, and how do they relate to the overall need for information to move the process forward?
8. Social outcomes	What are the social relationships (i.e., people communicating with each other) to move this database design ahead as part of the overall decision process? Did we make any enemies or new friends, or meet new colleagues that could be advantageous for carrying out the GIS project?

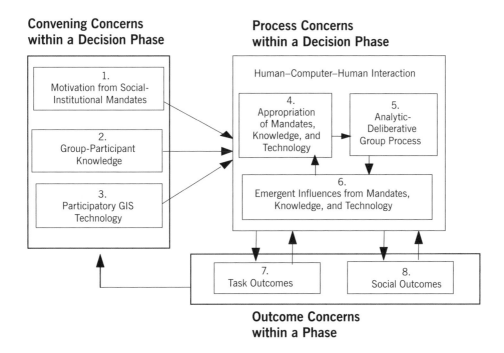

FIGURE 4.1. Relationships among eight constructs grouped by similar concern within a single decision support phase. Groupings are not to be interpreted as phases.

assessment for the representation modeling phase only (Table 4.2), and consider the convening, process, and outcome concerns, and the respective constructs within the representation modeling phase. This example elucidates convening, process, and outcomes *constructs* for the "phase" of problem representation; hence, there are convening, process, and outcome activities associated with this phase. The convening constructs are concerns involving motivation, groups participating, and technology being used. That *macro*-phase and *micro*-activity perspective makes EAST2 a flexible framework for complex problem solving. EAST became EAST2 when we further elaborated the framework and described its use in a number of decision situations. Any macro phase can be described in terms of any number of micro activities. That macro–micro recursion, that is repeating over again, is common on many processes.

The phase by construct level of assessment is useful for developing a more detailed understanding of a decision situation than was available at the phase level alone. When this level of assessment is shared among coworkers and/or participants, they become able to form a shared understanding about the overall decision situation. Unfortunately, in some parts of the process, we might feel that we know very little as yet; in other parts, we might feel we know way too much. To develop an even better understanding of the decision situation, particularly when GIS technology will be an instrumental part of the analytic–deliberative process, a more detailed level of assessment could be useful, and we could undertake a phase by construct by aspect assessment.

4.2.4 Decision Situation Assessment
by Phases–Constructs–Aspects

Before we jump into this level of assessment let us ask a question about our overall
understanding of the decision support process in regard to where we feel we need to
know more and where we feel we do not need to know more; that is, we deal with
aspects.

Each of the eight constructs can be described in terms of *aspects* of the decision
situation (Table 4.3); yes, a new word, but at least we did not use the same word, so that
everything would be confusing. When we identify and describe the aspects, we know
considerable detail about that portion of the decision support situation. For example,
consider any aspect of Table 4.3 and ask yourself what you know about the Green County
GIS project in regard to that aspect. Does it matter whether you know something about
that aspect of the decision support situation? Every aspect has appeared multiple times
in various publications that deal with some "real-world" decision situation (Jankowski
and Nyerges 2001a), which is why the decision situation assessment framework can
lead to a deeper and broader understanding of just about any decision situation. When
we take the framework to this level, and consider the relationships among constructs
and aspects, we are using the framework at a theoretical level of EAST (Nyerges and
Jankowski 1997).

One of the advantages of now having the ability to perform a decision situation
assessment at any level of detail is that an analyst can customize the assessment to meet
the needs of a particular circumstance. Let us take this approach and devise a Green
County decision situation assessment that uses all aspects of the situation, but for the
entire situation rather than for individual phases (see Table 4.4). What we have is a
general assessment at the aspect level; that is, we combined a Level 1 assessment and
a Level 4 assessment. Many of the assessments performed by GIS analysts using this
framework have done just that. As previously mentioned, a general assessment takes the
least amount of time, but the aspect assessment provides the most detail. This level of
assessment shows us the nature of the realism in the Green County GIS project, which
should lead the reader to be curious about how certain aspects might actually affect a
Green County GIS project. If the reader pursues this thought, Table 4.4 provides a great
way to orient oneself to what might matter in a real GIS project.

The Green County decision situation assessment presented in Table 4.4 provides
an analyst, and all involved in the decision situation, with a common interpretation, if
not understanding, with an overview of the wastewater facility site selection process. It
is only an overview, as we mentioned earlier; many phases (at least four in regard to the
GIS project) may be involved. Clearly, there is much at stake. Consequently, many stakeholder groups are likely to be involved in these and other complex decision problems.
A decision situation assessment strategy is one way to address the complexity without
getting overwhelmed.

Complex planning, improvement programming, and project implementation decision problems are not easily solved with simple solutions. If one could do so, then it

TABLE 4.3. Twenty-Five Aspects of EAST2: A Theory of GIS-Supported Participatory Decision Making

Constructs and aspects about convening a participatory situation

Construct 1: Social-institutional influence

1. Power and control
2. Subject domains, such as task purpose, content, and structure
3. Persons, groups, and/or organizations as convener of participants
4. Choosing the number, type, and diversity of participants
5. Rules and norms as social structures among participants

Construct 2: Group participant influence

6. Participants' expectations based on values, goals, issues, values, beliefs, and fairness
7. Participants' views/knowledge of the subject domain and each other
8. Participants' trust in the process
9. Participants' beliefs and feelings toward information technology

Construct 3: Participatory GIS influence

10. Place, time, and channel of communications
11. Availability of social–technical structures as information aids

Constructs and aspects about participatory process as social interaction

Construct 4: Appropriation

12. Appropriation of social-institutional influence
13. Appropriation of group participant influence
14. Appropriation of participatory GIS influence

Construct 5: Group process

15. Idea exchange as social interaction
16. Participatory task flow management
17. Behavior of participants toward each other

Construct 6: Emergent influence

18. Emergence of participatory GIS influence
19. Emergence of group participant influence
20. Emergence of social-institutional influence

Constructs and aspects about participatory outcomes

Construct 7: Task outcomes

21. Character of decision outcomes
22. Decision outcome and participant structuring dependence

Construct 8: Social outcomes

23. Opportunity for challenge of the outcome
24. Reproduction and temporality of group participant influence
25. Reproduction and temporality of social-institutional influence

Note. From Jankowski and Nyerges (2001a). Copyright 2001 by Taylor & Francis Ltd. Reprinted by permission.

TABLE 4.4. Decision Situation Assessment for Green County

3 components, 8 constructs, and 25 aspects	Project situation: Describe each aspect in sufficient detail.

Convening component

Construct 1: Social-institutional influence

1. Power and control

2. Subject domain as task purpose, content, and structure

3. Persons, groups, and/ or organizations as convener of participants

4. Choosing the number, type and diversity of participants

5. Rules and norms as social structures among participants

1. The mandate for a wastewater facility comes from a vote of the County Council. That vote is based on population growth information coming from the Planning and Public Works joint study effort. Other federal, state, and likely county regulations apply, particularly environmental protection. Various stakeholders have rights to clean water that are protected under county, state, and federal regulations.

2. The purpose of the task is to develop a GIS-based siting process for a wastewater facility that is defensible to all stakeholders. Although the County Council will be the main stakeholder, others will likely be involved at some time.

3. The County Council is the convener of the process, but the Public Works Department is the responsible unit for conducting the siting process.

4. A GIS analyst is performing the technical task, but a variety of other participants will evaluate information the analyst derives. Participants could range from technical specialists through managers to lay participants not versed in the use of computer technology and decision support tools.

5. Whereas the County Council participants come from a variety of backgrounds, they are the elected officials. However, a residents' committee was selected and formed to lead/speak for various community interests. All participants are comfortable with defending and promoting their positions/interests, and not be intimidated by the process.

Construct 2: Group participant influence

6. Participants' expectations based on values, goals, issues, beliefs, and fairness

7. Participants' views/ knowledge of the subject domain and each other

8. Participants' trust in the process

9. Participants' beliefs and feelings toward information technology

6. Each participant expects to take part in a process that will fully address the rights and concerns of the group he or she represents, including the elected officials, the technical specialists, and the residents committee.

7. The knowledge of each participant is focused primarily on the activities and concerns of the group he or she represents. Part of a participatory process is to educate one group about the concerns of other groups.

8. In general, participants are anticipated to be wary of governmental interaction with water resources issues, and protective of their respective domains of interest.

9. Many participants are familiar with the application of information technology to resource issues, but many are still wary of its use. The GIS analysts in the process will be called upon to explain how they derived the information during their work process.

(cont.)

TABLE 4.4. *(cont.)*

3 components, 8 constructs, and 25 aspects	Project situation: Describe each aspect in sufficient detail.

Construct 3: Participatory GIS influence

| 10. Place, time, and channel of communications | 10. GIS analysts will convene at GIS workstations that are connected to local and wide area networks. Are the GIS analysts the only participants who will have access to data files? Will any of the County Council staff have access to data and maps? What about the residents? |
| 11. Availability of social-technical structures as information aids | 11. Analysts will provide tables, maps, and charts as necessary to the GIS, taking advantage of GIS-based decision support capabilities/computers connected to a local area network. |

Process component

Construct 4: Appropriation

12. Appropriation of social-institutional influence	12. A GIS analyst comes to the siting process with knowledge of the County Council's interest in siting a wastewater facility. Hence, the analyst is motivated to work in line with that social-institutional influence. Residents of Green County, however, might come to the process with a variety of influences that actually have reasonable (if not better) standing in the community.
13. Appropriation of group participant influence	13. A GIS analyst's responsibility is to provide a recommendation in line with the stakeholder influences according to the Director of Public Works. A GIS analyst is a professional, and should adopt an ethical stance in devising recommendations. A GIS analyst needs to pay attention to all comments made in relation to the appropriation of various social-institutional influences, because those influences are the basis of the criteria used in the siting process.
14. Appropriation of participatory GIS influence	14. A GIS analyst employed/contracted by Green County will have access to the GIS technology to perform the siting task. This aspect should be described in more detail, perhaps for each of the phases of the project, to make sure the appropriate software is available. What if the Council wanted residents to have access to maps? What user interface would be appropriate for such access? The same as that used by the analysts?

Construct 5: Group process

15. Idea exchange as social interaction	15. Idea exchange is conducted verbally and in relation to maps, tables, and charts presented. The exchange is enabled and constrained by the communication channels identified and appropriated. If the convener did not consider the availability of widespread area network communications for participants, then such broad-based communications with a variety of residents will not likely occur.
16. Participatory task flow administration	16. The decision agenda and the GIS workflow are selected by convener and participants, respectively. Each has expected outcomes, but those outcomes are dependent on implementing the processes. Analysts' workflow is dependent on what they have learned about workflow processes.
17. Behavior of participants toward each other	17. Participants' behavior toward each other depends on the social mores they have learned and how they put these into practice.

(cont.)

TABLE 4.4. *(cont.)*

3 components, 8 constructs, and 25 aspects	Project situation: Describe each aspect in sufficient detail.

Construct 6: Emergent influence

18. Emergence of GIS influence

18. The influence of new GIS capabilities in the workflow process is dependent on the curiosity and exploration of analysts.

19. Emergence of group participant influence

19. Group participants' influence is anticipated to emerge in a variety of groups depending on how they interpret the activities in the GIS workflow.

20. Emergence of social-institutional influence

20. In this situation, the decision maker is the County Council, as defined by the incorporation charter of the city. However, in many cities, the Mayor actually has responsibility for capital projects; thus, he or she could be the decision maker. However, the County Council would then have to approve (invoking a checks and balances process).

Outcome component

Construct 7: Task outcomes

21. Character of decision outcomes

21. The process is intended to result in the preparation of an equitable, effective, and efficient recommendation of sites (or site) that can be defensible by virtue of the process used. The recommendation takes the form of a report.

22. Decision outcome and participant structuring dependence

22. The primary task of the analyst is to prepare and present a recommendation to the Director of Public Works, who then formally provides the report to the County Council.

Construct 8: Social outcomes

23. Opportunity for challenge of the outcome

23. The County Council decides whether to release the report to residents for comment. A public meeting might be held. If state and/or federal funds are involved in the project, most likely a public meeting will occur, providing residents with opportunity for comment.

24. Reproduction and temporality of group participant influence

24. From time to time, County Council membership changes, public works departments change, and GIS analysts come and go. Those who have taken part in the effort do have an influence; thus, the stability of the recommendation is based on those influences.

25. Reproduction and temporality of social-institutional influence

25. Social-institutional influences are long term and relatively stable. Sometimes, legal issues arise. For example, the Residents' Committee, or perhaps some other coalition of community residents, could choose to request intervention by state authority or sue the city due to inadequate "due process." For this reason, more and more organizations are inviting broad-based participation at the beginning that continues throughout the decision process. This turns out to be a lot less expensive in terms of time and money.

would not take a lot of people to address such problems, and there would not be so many complaints about lack of _due process._ The decision situation assessment framework is complex because many contingent aspects need be considered. Nonetheless, let us take a step back from the long tables just introduced in this section, and present a summary that includes the three categories of constructs, the eight constructs, and the 25 aspects (see Figure 4.2). Once a GIS analyst understands the three categories, then the eight constructs are not difficult to understand. In the same way, once an analyst understands the eight constructs, then the 25 aspects are not difficult to associate with those constructs; it is a matter of scaffolding knowledge.

At first, most people are overwhelmed by the level of detailed information in the framework. But now that we have covered the assessment levels of the framework, each level should be a bit more understandable. Analysts who have applied the framework to complex decision situations find the results quite informative—more than they would have expected (Drew 2002; Tuthill 2002; and Miles 2004). All have said, "My goodness, I now know things that I would have never understood without doing the assessment, and it was not very hard either—just step by step." However, we reiterate that the most important point is that the framework can be applied at various levels of detail for the general situation, or for one or more of the phases in a workflow process. Once an analyst decides about workflow, then he or she can make choices about what information

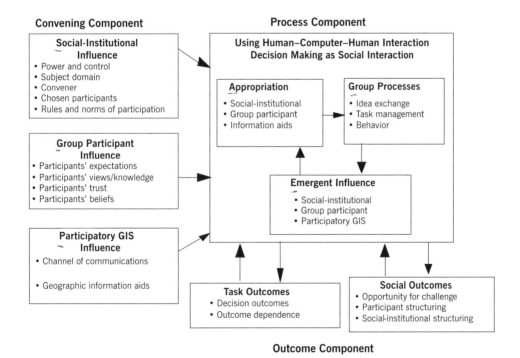

FIGURE 4.2. Summary diagram for 25 aspects (depicted as bullet phrases) of EAST2 framework.

from the framework is necessary or useful, depending on the project circumstances. The reader can choose what to know and when to know it about his or her own decision situations.

4.3 Summary

As mentioned in previous chapters, and expanded upon in this chapter, GIS is touted as a decision support system. This chapter presents many details about how and why GIS can assist with decision support. We expanded on the details by first characterizing geospatial decision problems in terms of two basic dimensions, ranging from simple to complex, then from content to context. Solving complex decision problems motivates the use of GIS. More complex decision problems are likely to require more complex workflows.

We outlined several approaches for performing decision situation assessment in terms of level of detail. A decision situation assessment provides GIS analysts a framework with which to think through the needs of GIS decision support problems—not only from an information perspective but also from a broad-based organizational perspective. In that section we described four levels of detail for assessments. A general assessment level provides insight an at overview level, focusing on convening, process, and outcomes of the decision situation. In slightly more detail is an assessment by phase to describe the overview categories, but for each of the phases involved. The third level, a construct assessment by phase, details constructs for each of the phases. The fourth level, a construct–aspect assessment by phase, details aspects of all constructs by phase in a decision problem.

Readers should take the detail with guarded concern and use only the level of detail to unpack the complexity of decision situations as needed. If a GIS analyst continues to address similar decision problems over and over again, then there will be less need for an assessment in advance of addressing a decision situation. We, and our students, continue to find the framework useful because of the variety of geospatial decision problems to which GIS is being applied. We encourage readers to be creative with its use; it is not a rigid framework, but a flexible one that can be customized to the situation at hand. We use the decision assessment framework in varying detail within later chapters to describe several GIS projects.

Before we present actual analyses that others have done, the reader should have a better understanding of GIS databases and analysis. The core of how we can build database designs is incorporated into the idea of *data models*, the topic of Chapter 5. We use various data models in the design particular databases. The design of a particular database is what we call a *database model*. Implementation of a database model results in a *database*—the data we store to provide decision support for a particular GIS project.

[handwritten margin notes: GIS projects - content, structure, process, context are... (considerable problems)]

4.4 Review Questions

1. Why does it matter whether we understand the difference between a GIS project and a capital improvement project when talking with people other than those familiar with our GIS project?

2. What is the relationship between the three perspectives of GIS embedded within the definition of GIS provided in Chapter 1 and the following three topics: (a) decision problems, (b) components of software, and (c) the phases of a GIS project?

3. What are the dimensions of the decision problem framework?

4. How might we characterize the difference between closed systems decision problems and open systems decision problems in relation to content, structure, process, and context characteristics of such problems?

5. What is decision situation assessment; and how do we use it to improve our understanding of GIS-based workflow?

6. What is a general assessment for a decision situation assessment?

7. What is a by-phase assessment?

8. What is a by-phase and construct assessment?

9. What is a by-phase-construct–aspect assessment?

10. How would you customize a decision situation assessment for the Green County GIS project?

PART II

Fundamentals of GIS Data and Analysis for Decision Support

CHAPTER 5

Making Choices about Geospatial Database Development

Database Design = data modeling) — imple.

The workflow processes presented in Chapter 3 indicate that database development is one of the most important activities in GIS work. Data modeling, or what is commonly called *database design*, is a beginning step in database development. Database implementation follows data modeling for database design. The implementation of a database design in part depends on what software is used to perform the implementation. Even database management software must be designed with some ideas about what kinds of features of the world to represent in a database. No database management software can implement *all* feature representations and needs. Such software would always be in design mode. Limits and constraints (i.e., general nature of GIS applications to be performed) exist for all software.

To gain a better sense of data modeling for database design we first distinguish between data and information. If there were no distinction, then software would be very difficult to develop, and GIS databases would not be as useful. We then differentiate between data models and database models as a transition into how databases relate to software used to manipulate data. We then present a general database design process that can be used to design geodatabases—the newest and most sophisticated types of GIS databases currently in use.

5.1 Data, Information, Evidence, and Knowledge: A Comparison

Because data modeling deals with classes of data, it is really more about information categories. Some might even say that the data classes are about knowledge, because the categories often become the basis of how we think about GIS data representations. To gain a sense of data modeling, let us define the terms *data*, *information*, *evidence*, and *knowledge*. Those terms are not always as well understood in common practice as in everyday language. Defining the terms provides a clearer sense of their differences and relatedness to help with data modeling, as well as a basis for understanding how information products relate to knowledge in a broader sense. In a GIScience and systems context, Longley et al. (2005) have written about the relationships among those terms. The definitions provided below are based on interpretations of Longley et al., integrated with Sayer's (1984) geographic treatment of epistemology, because GIS analysts work in contexts involving many perspectives.

- *Data*—are raw observations (e.g., a measurement) of some reality, whether past, current, or future, in a shared understanding of an organizational context. We typically value what we measure, and we measure what we value. One might ask: What is important enough to warrant expending human resources to get data?

- *Information*—is data placed in a context for use that has meaning about a world we share. Geographic information is a fundamental basis of decision making; hence, information needs to be transparent in groups, if people are to share an understanding about a situation.

- *Evidence*—is information that makes sense (perhaps corroborated); hence, it is something we can use to make reasoned thought (argument) about the world. All professionals (whether they be doctors, lawyers, scientists, GIS analysts, etc.) use evidence as a matter of routine in their professions to establish shared valid information in the professional community. Credible information is the basis of evidence. How we interpret evidence shapes how we gain knowledge. When we triangulate evidence we understand how multiple sources lead to robust knowledge development as the evidence reinforces or contradicts what we come to know.

- *Knowledge*—is an assemblage of synthesizing, enduring, credible, and corroborated evidence. Knowledge enables us to interpret the world through new information and, of course, data. We use knowledge about circumstances to interpret information and decide whether we have gained new insight. It is what we use to determine whether information and/or data are useful or not.

The purpose of elucidating the levels of knowing is to provide readers with the perspective that GIS is not just about data and databases, but that it extends through higher levels of knowing. Given all that has been written and researched about these relationships, most people would say there are many ways to interpret each of the data, information, evidence, and knowledge steps.

With these distinctions in mind, in this chapter we present a framework for understanding the choices to be made in data modeling. *Data modeling* is a process of creating database designs. Both data models and database models are used to create and implement database designs. We can differentiate data models and database models in terms of the level of abstraction in a data modeling language. A database design process creates several levels of database descriptions, some of which are oriented for human communication, whereas others are oriented to computer-based computation. *Conceptual*, *logical*, and *physical* are terms that have been used to differentiate levels of data modeling abstraction. A database model focuses on the data constructs only. A data model includes not only constructs but also operations and integrity constraints on those constructs and operations. James Martin (1976) and Jeffrey Ullman (1980) popularized the term *data model* to refer only to the constructs. However, Edgar Codd (1970), the inventor of the relational data model, included the operations and integrity constraints, because not all data relationships can be stored within a database. Some relationships must be processed through operations that are constrained by rules.

We form data models as languages using basic constructs, operations, and constraint rules. Such languages provide us with the capabilities to develop specific database designs. The database designs we create are like a type of story about the world; that is, we limit ourselves to certain *constructs* (categories for data), together with some potential operations on data, to tell a story through a template for a database representation. Thus, we include some feature categories of points, lines, and polygons, but we exclude others. It depends on what we want to do (the kind of data analysis or display we might perform) with the story.

The constructs of a data model and how we put them to use are often referred to as *metadata*. Metadata are information about data. They describe the particular constructs of a data model and how we make use of them. Data category definitions need to be meaningful interpretations to be able to model data. However, there are at least three levels of metadata (data construct descriptions and meaningful interpretations) in a data modeling context. So let us explore these three levels for each of the data model and database model interpretations in the following sections.

5.2 Data Models: The Core of GIS Data Management

Before we get started, consider for yourself which of the three levels of conceptual, logical, and physical is *more abstract* (i.e., which level is more general or more concrete for you). The *conceptual level* is about meaning and interpretation of data categories. The *physical level* is about the bits and bytes of storing the data. For some, the conceptual model is more abstract, whereas for others, the physical model is more abstract. From the viewpoint of a database design specialist, the move from general to specific detail proceeds from the conceptual through the logical, and on to the physical. With that in mind, we tackle the levels in the order of abstractness.

5.2.1 Conceptual Data Models

A *conceptual data model* organizes and communicates the meaning of data categories in terms of object (entity) classes, attributes and (potential) relationships. We can use a natural language, such as the English language, or a database diagramming language to express the main ideas in a database design. Our choice really depends on the people participating in the design. Natural language has the advantage of being more easily understood by more people. However, natural language has limitations, in that it is not often as clear or precise, because it is unconstrained in its semantic and syntax expression. People express themselves with whatever *constructs* (nouns) and *operators* (verbs) they have learned as part of life experience. A diagramming language, for example, an entity–relationship language, is a "stylized" language; that is, it adopts certain conventions for expression. As such, the expressions tend to be clearer than natural language. However, people need to learn such a language, like any language, to be proficient in expression.

The following English language statements lead to comparable expressions in an entity–relationship (ER) language (see Figure 5.1). The language is the oldest conceptual database design language in use, popularized by Peter Chen (1976); the ER language was actually his dissertation and was later published. It quickly caught on because of its simplicity, and also because data management technology was growing in importance and receiving attention in the information technology world.

- The facilities will be located on land parcels, with compatible land use.
- Streams/rivers should be far enough away from the facility.
- A street network will service the facility.

In a natural language, nouns are often the data categories. The expressions often provide information other than categories, such as surrounding features. In English, the categories could be either a singular or plural form as a natural outcome of usage. In an ER expression, by convention, the data categories are singular nouns. Nonetheless, there is a correspondence between the English and the ER expressions; that is, nouns are the focus of data categories.

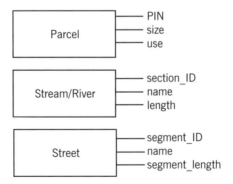

FIGURE 5.1. Simplified entity–relationship diagram showing only entity classes (boxes) with attributes (on the right side after lines), and showing no relationships.

Because the ER language is part graphic and part English, we can also use a purely graphical language to depict the differences among geodata entity types, particularly in consideration of the spatial aspect of geodata.

Remember that you learned earlier three special aspects to data models: the constructs, operations (that establish relationships), and integrity/validity constraints (rules). As the first aspect, spatial data constructs in geospatial data models comprise geospatial object classes, also called *data construct types* by some people (Figure 5.2).

The geospatial data construct types differ from each other due to geometric dimensionality and topological relationships stored (or not) as part of the data constructs. Basic geometry is given by dimensionality. Data construct types of a 0-, 1-, 2-dimensional character are shown in Figure 5.2. *Points* are 0-dimensional mathematical object constructs defined in terms of a single coordinate (or *triordinate space*). However, *shape* (e.g., shape of a polygon) within a dimensionality is a natural outcome of the storage of specific coordinates. The coordinates are an outcome of the measurements of locational relationships.

Some data models contain only geometric geospatial constructs (i.e., just points, lines, and polygons), represented through use of coordinates. No spatial relationships (called *topology*) are stored in the data model constructs. These relationships have to be computed if they are to be known. This leads us to the second aspect, which deals with operations.

The second major aspect of data models concerns *operations* (i.e., relationships among constructs). Operations are a way of deriving relationships. *Topology is* the study of three types of relationships—connectedness, adjacency, and containment—among objects embedded in a surface. Topology can be stored implicitly or explicitly in a data model. Implicit storage stems from using simple constructs in a representation, such as a construct (e.g., a cell in a grid structure or a pixel in a raster structure). The cells or pixels in their respective data structures are each the same size. Thus, the relationship termed *adjacency*, meaning "next to," can be assumed/computed based on cell size. Connectedness, as a relationship derived from the adjacency, can be determined by taking a data structure walk from one grid cell to the next. Adjacency and connectedness derive from the same "next to" relationship.

When geospatial objects are not the same size, adjacency must be stored. The most primitive topological object is called a *node*. The relationship that occurs when two nodes are connected is called a *link*. Vector data constructs, such as the nodes and links in Figure 5.2, must have explicitly stored relationships to express adjacency, connectedness, and containment to compose a topological vector data model.

A third major aspect of conceptual data models is the types of *rules* that assist in constraining operations on data elements. One important type of rule, a *validity rule*, maintains the valid character of data. No data should be stored in a database that does not conform to the particular construct type that is being manipulated at the time. Another kind of validity rule is how relationships among data elements are established, for example, object-oriented data models that can represent the logical connectedness between features, such as storm sewer pipes. In such a data model, each segment in

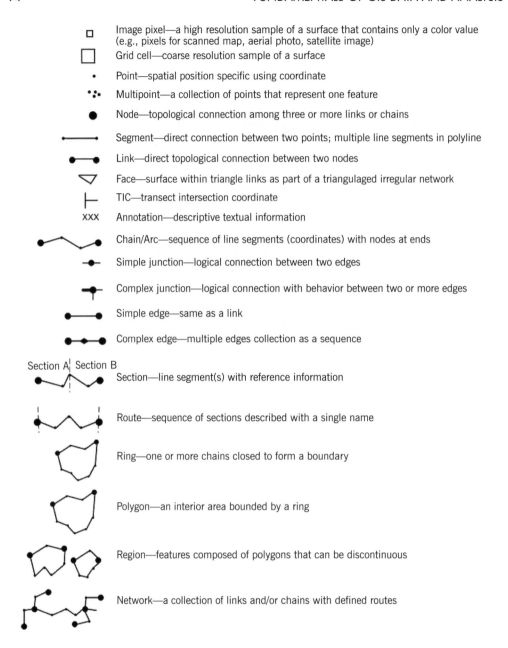

□   Image pixel—a high resolution sample of a surface that contains only a color value (e.g., pixels for scanned map, aerial photo, satellite image)

▢   Grid cell—coarse resolution sample of a surface

•   Point—spatial position specific using coordinate

•ᐧ•   Multipoint—a collection of points that represent one feature

●   Node—topological connection among three or more links or chains

●———•   Segment—direct connection between two points; multiple line segments in polyline

●———●   Link—direct topological connection between two nodes

▽   Face—surface within triangle links as part of a triangulaged irregular network

├—   TIC—transect intersection coordinate

xxx   Annotation—descriptive textual information

●⌒⌒●   Chain/Arc—sequence of line segments (coordinates) with nodes at ends

—●—   Simple junction—logical connection between two edges

—●—   Complex junction—logical connection with behavior between two or more edges

●———●   Simple edge—same as a link

●—●—●   Complex edge—multiple edges collection as a sequence

Section A⎸ Section B
●⌒⌒●   Section—line segment(s) with reference information

⌇⌒⌒⌇   Route—sequence of sections described with a single name

⬡   Ring—one or more chains closed to form a boundary

⬡   Polygon—an interior area bounded by a ring

⬡⬠   Region—features composed of polygons that can be discontinuous

⌇⌒⌇   Network—a collection of links and/or chains with defined routes

FIGURE 5.2. Common geospatial data construct types for raster and vector data models. Data from National Institutes for Standards and Technology (1994) and Zeiler (1999).

an object class called "storm sewer pipe" is to be "connected" to only one other "storm sewer pipe," unless a valve or junction occurs. Then, three pipes can be connected. In addition, storm sewer pipes can only be connected to sanitary sewer pipes if a valve occurs to connect them.

Why choose to use one conceptual data model rather than another for any particular representation problem? Each has its special character for depicting certain aspects about the geospatial data design. None is particularly superior for all situations.

5.2.2 Logical Data Models

Logical data models are developed as a result of including certain geospatial data constructs in the software design of the data model in particular ways. Choosing particular ways of representing data both *enables* and *constrains* us to certain data processing approaches. Several GIS software vendors offer various approaches to logical data models; it is what distinguishes one solution from another.

Several GIS software vendors provide great solutions for GIS computing directed at various market segments. As such, they have a tremendous assortment of GIS software from which to choose. Among the vendors and products are the Environmental Systems Research Institute (ESRI) with ArcGIS, Unisys with System 9, Caliper Corporation with TransCAD, GE Energy with Smallworld GIS, and MapInfo Corporation with MapInfo.

As mentioned previously, a data model comprises three components: data constructs, operations, and validity rules. The combination of these three components is what makes data models different from one another. However, the data construct component is commonly viewed as the most fundamental information, because without data constructs, there would be no data; hence, there would be no need to perform data processing. The different vendors offer different nuances in their data models.

Because there are too many data models to cover in the space of this textbook, we look at the most popular (best selling) among them, the ArcGIS data models from ESRI (2006), which has been developing and distributing GIS software for over 30 years. ESRI is a world leader in GIS software in terms of number of installations, which is why we use their data models as a basis for this discussion of logical data models. Because the installed customer base is so large, legacy issues must be addressed (i.e., installed software of older database systems). There is a tremendous challenge to develop new approaches to geospatial data organization, while simultaneously maintaining an installed legacy base. That is why current conversion programs and vendors are doing good business.

ArcGIS logical data models are the raster or image/grid data model, the triangulated irregular network (TIN) data model, shapefile data model, coverage data model, and the geodatabase data model (ESRI 2003). The TIN and the grid data models are often used to represent continuous surfaces. The shapefile, coverage, and geodatabase data models are used for storing points, lines, and areas that represent mostly discrete features. Early on, many researchers distinguished the two types as raster and vector data models. Later on, others referred to the difference as objects and fields (Cova and Goodchild

2002). Fundamental differences in surfaces/fields and objects/features lead to a difference in the design and implementation of data models described in terms of the three components: data constructs, operations, and validity constraints. We pick up from the conceptual data constructs of the previous section and show how implementation of these constructs has led to organization of a data model in a particular way. First we treat all data constructs. We then address the operations for each and, finally, the validity constraints.

5.2.2.1 Data Constructs of Six ESRI Data Models

Differentiation of the data models in terms of data construct types is the most well-known distinction among them. In Table 5.1 the data models are listed left to right, roughly in terms of complexity, although these are only rough approximations. The constructs provide a comparison of basic structure among the six data models.

The raster image and grid, and the TIN data models are used for representing surfaces. There are differences among them in terms of the spatial data construct types

TABLE 5.1. Spatial Data Construct Types Associated with Data Models

Spatial data construct type	Logical data models					
	Raster data models		Vector data models			
	Image	Grid	TIN	Shapefile	Coverage	Geodatabase
Image cell	×					×
Grid cell		×				×
Point				×	×	×
Multipoint				×		×
Node			×			
Segment/polyline				×	×	×
Link			×			
Chain/arc					×	
Face			×			
Tic					×	
Annotation				×	×	×
Simple junction						×
Simple edge						×
Complex junction						×
Complex edge						×
Section					×	
Route					×	
Ring				×		×
Polygon				×	×	×
Region					×	
Network					×	×

used to represent surfaces (see Table 5.1). The *grid* provides for a coarse resolution of sampling data points that are commonly arrayed as a regular spacing. The grid data model is meant to represent elevation surface. It is also known as a digital elevation model (DEM), because that is the topical area in which it received considerable use. Rather than focusing on points of information content that stand out for special reasons, the grid data model samples points at regular intervals across a surface. As such, it uses considerable data, because it is an exhaustive sampling technique. Because topological relations among grid points are implicit in the grid, geometric computations are very quick.

The raster data model became popular when satellite imagery was introduced. The density of the regular spacing of points became quite high, and a variety of software has been developed to store and manipulate images. As such, the raster data model includes both *image* and *grid cell* data models. The pixel points are commonly regularly spaced, although, theoretically, they do not have to be regularly spaced. Data processing of regularly spaced points is much easier than that for irregularly spaced points (samples).

The TIN data model is meant to represent elevation surface, but any surface can be modeled by a TIN. A TIN takes advantage of known feature information to compose the surface representation; thus, it is parsimonious with data. Peaks, pits, passes, ridges, and valleys can be included in the model as high information content locations, also called *critical points*. They are critical for capturing the lows and the high elevation points on a surface. Topological relations among vertices are explicit in the TIN, and use nodes (peaks and pits) and links (valleys, passes, ridges, etc.) to represent the surface. Because three points define a plane, the surface planes bend easily along the edges of the planes to characterize a surface. Note the peaks, pits, passes, ridges, and valleys along the edges that can be used to trace a path.

The shapefile data model contains features with no topological relationships; it contains geometry only (see Table 5.1). The points in a shapefile are commonly irregularly spaced, representing point-like features in the world. They are taken individually to be meaningful. This is perhaps the major difference in points within the three surface data models described previously and the shapefile, coverage, and geodatabase data models. The multipoint spatial construct can be used to represent a cluster of points, such as a given set of soil samples taken in a field at one point in time. That specific set is retrieved with a single ID, rather than an ID for every point. The line within the shapefile data model can be line segments (straight line from point to point), circular arcs (parameterized by a radius, and start and stop points), and Bézier splines (multiple curves to fit a series of points).

The coverage data model had been the mainstay of ArcInfo software for almost 20 years. It is also called the *georelational data model*, comprising spatial and attributes data objects. The coverage includes feature classes with topological relationships within each class (no topology between layers; e.g., a river network would not be part of a transportation network if the transportation network is a highway network; see Table 5.1). The primary objects are points, arcs, nodes, and polygons within coverages. Topological arcs and nontopological arcs (polylines) are possible. Arcs close (start and end coordinates match) to form a ring (boundary) of a polygon. Secondary objects are tics, links, and annotations. Transect intersection coordinates (tics) are used to provide the spatial reference.

The geodatabase data model is the most recent of the ArcGIS data models. It contains objects that provide functional "logic," temporal logic, as well as topo (surface) logic relationships. Logical relationships with constraints provide the most flexibility for modeling feature structure and process (see Table 5.1 for geodatabase constructs). The feature classes can be collected into similarly themed structure in what are called *feature datasets*. Topological relationships can span feature classes when included in a feature dataset. The base features include categories for generic feature and custom feature classes. The generic feature classes (feature specification in general) include point, multipoint, line (line segment, circular arc, Bézier spline), simple junction, complex junction, simple edge, complex edge, and custom feature classes.

One fundamental question is why use one data model rather than another? Let us consider some of the advantages and disadvantages of the data models in relation to each other (Zeiler 1999). In the geodatabase model, the spatial data and the attribute data are at the same level of precedence; that is, either one can be stored, followed by the other. However, in the coverage data model, the spatial data geometry must be stored first, then the attribute data. The shapefile data model must also store a geometry first (point, polyline, polygon); then the attribute can be stored. Temporal data in the geodatabase is stored as an attribute; similarly it is in the shapefile and the coverage, but with its own "special domain" of operations. The geodatabase data model was developed to provide for built-in behaviors—featured ways of acting (implemented through rules) that can be stored with data. In contrast with the coverage and the shapefile data models, the geodatabase manager performs data management using a single database manager as a relational object rather than as file management, as in the shapefile, and file management and database management, as in the coverage data model. Large geodatabases do not need to be *tiled* (squares of physically managed space) using a file manager, as in the coverage data model. There is no opportunity for very large database management in the shapefile model. In addition, the geodatabase environment allows for customized features such as transformers, parcels, and pipes (not geometry defined, but attribute defined).

5.2.2.2 *Relationships Underlying the Operations of Six ESRI Data Models*

The second major component of a data model is the set of operations that can be applied against the data constructs for that model. There are four basic types of operations for data management: create (store), retrieve, update, and delete. Of course, what actually gets created, retrieved, updated, and deleted is based on what data model constructs are being manipulated. All of the data models contain many specialized operations that make sense only for that data model, because of the inherent information stored within the structure of the data constructs. We have addressed these operations in more detail in Chapter 6 when considering analysis, but let us characterize the major differences

TABLE 5.2. Spatial, Logical, and Temporal Relationships Underlying Operation Activity

Relationship	Data models					
	Image	Grid	TIN	Shapefile	Coverage	Geodatabase
Spatial distance and geometry	Derived	Derived	Derived	Derived	Derived	Derived
Spatial topological	Implicit	Implicit	Explicit	Derived	Explicit	Explicit
Connectedness	Implicit	Implicit	Explicit	Derived	Explicit	Explicit
Adjacency	Implicit	Implicit	Explicit	Derived	Explicit	Explicit
Containment	Derived	Derived	Explicit	Derived	Explicit	Explicit
Function logical	Derived	Derived	Derived	Derived	Derived	Rules stored
Temporal logical	Derived	Derived	Derived	Derived	Derived	Rules stored

Note. Implicit—stored as part of the geometry, easily derived; explicit—stored within a field, easily processed; derived—information computable, but time consuming; none—cannot be processed from available information.

among the data models by examining the spatial, logical, and temporal relationships inherent within the models (see Table 5.2).

Distance operations are fundamental distinguishing characteristics of spatial analysis in a GIS. Thus, distance is derived in all data models. The raster and grid data models contain a single spatial primitive—the cell/pixel. As such, the spatial topological relationships are *implicitly stored* within the data model based on the row and column cell position, making the spatial topological operations very easy and powerful. Topological relationships are explicitly stored in TIN, coverage, and geodatabase models. However, because shapefile data model spatial data construct types are all geometric, they require computation of topological relationships, which is why that category is labeled *derived*. No logical and/or temporal information is inherent in the data models except for the geodatabase model; thus, such information can be computed from attribute storage using scripting language software. A scripting language is like a high-level programming language (e.g., Visual Basic Scripting). The TIN, coverage, and geodatabase data models are the most sophisticated in terms of storing relationships. These data models support spatial topological, functional logical, and temporal logical operations more flexibly than all others. The newest data model is the geodatabase data model. As such, the functional relationships can be stored as customizable rules rather than requiring use of a scripting language to generate the relationships.

5.2.2.3 Validity Rules of Six ESRI Data Models

The third component of a data model is the set of validity rules that prohibit the operations from creating erroneous data content. Validity rules operate at the level of attribute field, keeping data content within a range of acceptable values (e.g., coordinates that should be within a particular quadrant of geographic space, or land use codes that must match the allowable zoning regulations). For example, if we are concerned about zoning within a quadrant of a city, then the land use code must be of a particular data value or set of data values: "residential" and/or "commercial" and/or "mixed."

Because the raster, grid, TIN, shapefile, and coverage data models were brought into commercial use before GIS software vendors understood the usefulness of integrity rules, they are not explicitly included in the data models. However, the geodatabase model, the newest data model, contains a variety of integrity rules, and functional integrity rules can be developed. For example, a data model can implement a set of rules whereby the kinds and/or sizes of a valve must fit a specified range of kind/sizes to connect water pipes on either side of a valve. Such rules can be considered part of the logical data model level but are implemented at the physical data model level, which we discuss next.

5.2.3 Physical Data Models

A physical data model implements a logical data model. Data-type implementation and the data-type field indexing are specified at the physical data model level. *Data type* refers to the format of the data. All of the data fields must have a clear data format specification as to how data are actually to be stored. Potential data types are listed in Table 5.3. Data indices support fast retrievals of data by presorting the data and establishing ways to use those sorts to look at only portions of the data when looking for a particular data element. Specifying data formatting and indexing details helps the database design perform well when transformed into an actual database.

Unified Modeling Language (UML) static diagrams within the Microsoft Office Visio software application use a table with tabs to specify data types as part of defining class properties (see Figure 5.3). Once the pointer in the categories of Figure 5.3 is set to attributes, the data types in the attribute portion of the window are set through a pulldown window under the "type" heading.

As a performance enhancement, indexing of fields can be added to the physical schema. Because databases commonly get very large, indices are added to the schema to improve data retrieval speeds. For example, R-trees (short for region trees) or quad trees (that subdivide into quadrants) are very popular means of partitioning a coordinate

---

TABLE 5.3. Data Types

- Numerical
  Integer—positive or negative whole number, usually 32 bits
  Long integer—positive or negative whole number, usually 64 bits
  Real (floating point)—single-precision decimal number
  Double (floating point)—double-precision decision number
- Character (text string)—alpha-numerical characters
- Binary—numbers stored as 0 or 1 expression
- Blob/image—scanned raster data of usually very large size
- Geometry shape (Figure 5.2 lists all shapes)

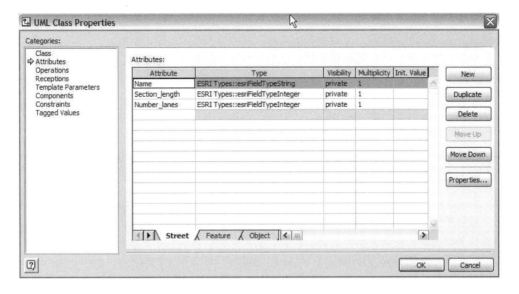

FIGURE 5.3. Physical schema specification for data types depicted using MS Office Visio UML class properties.

space without adding tremendous overhead for storage. These trees are logical organizations rather than physical organizations of data but are part of the physical schema, because they generate information at a very detailed level. The increase in speed of access to data for searching is well worth what little extra space is required.

Another consideration is where to store data on disk. It makes sense to keep data that are commonly retrieved with a similar time span near one another on the physical hard disk. The so-called "disk arm" does not need to move as much. This enhances the performance of retrieval for very large datasets but would not be very noticeable for smaller datasets. When a hard disk is not defragmented periodically, data are stored in many different physical locations, taking longer to retrieve.

5.3 Database Design Activity for Green County Functional Planning

A database model (using a particular data model schema) is an expression of a collection of object classes (entities), attributes, and relationships for a particular subject context (e.g., land resources, transportation resources, or water resources). Even more accurately, that context might be an application, or set of applications, for a particular topical domain of information, such as a transportation improvement programming or hydrological planning situation, in which the decision situation matters considerably.

A database model can be expressed at each of the three levels of abstraction—conceptual, logical, and physical—as described previously. These are called *levels of database abstraction*, because we choose to select (abstract) certain salient aspects of a database design. Different data models (languages), as presented in the previous sections, are used to create the database models. Representation in a database model depends on aspects of the world that need to be modeled and the choice of a data model to implement that representation.

What ESRI refers to as data models at its technical support website for data modeling are what we call *database models* at the conceptual level of design, with several logical characteristics (see ESRI 2006). The database models can be used to "jump start" GIS geodatabase designs. The descriptions of feature classes and attributes as a conceptual database design can be translated into logical database models (the choice of shapefile or geodatabases), then into physical database models. Having the opportunity to fine-tune the storage retrieval and access of the database is accomplished through the physical data model, as implemented within particular data management software for a particular type of operating system.

In review, the importance of the *conceptual* level is the name and meaning of the data categories (feature classes). The importance of the *logical* level is the translation of that meaning (we could also say *structure* of meaning) into several attributes for measurement (i.e., potential computable form) in a database management software system. The importance of the *physical* level is the actual storage of the measurements and the performance of the particular database in terms of how data are stored and retrieved from the disk.

The steps below outline a geodatabase design process adapted from Arctur and Zeiler's (2004) *Designing Geodatabases*, with some additions to complete the data modeling process set within the context of the Green County project we introduced in Chapter 3. The process includes conceptual, logical, and physical design phases. Each phase ends in the creation of a product called a *database model* (i.e., a structural representation of some portion of the world at an appropriate level of abstraction). The schema is the most visual part of that database model design.

As with any database design, the schema design is rather time consuming. As we mentioned earlier, a *schema* is a table structure in a relational model–oriented design. It is very important that data analysts understand how data are organized and, in particular, how to create nonredundant data expressions (also called *normalizing*) when designing a database. There are four approaches to building geodatabase schemas in ArcCatalog 9.x:

1. Create with ArcCatalog wizards.
   - Build tables in ArcCatalog >> right click >> new object.
2. Import existing data (and the existing schema).
   - Right click the database and import an object. You can also export from the object to the database.
3. Create a schema with computer-aided software engineering (CASE) tools.
   - Use Microsoft Visio or similar software for development of UML.
4. Create a schema in the geoprocessing framework.
   - Use ArcToolbox geoprocessing to create objects.

**TABLE 5.4. Geodatabase Database Design Process
as Data Modeling**

Conceptual design of a database model
- Identify the information products or the research question to be addressed.
- Identify the key thematic layers and feature classes.
- Detail the feature class(es).
- Group representations into datasets.

Logical design of a database model
- Define attribute database structure and behavior for feature classes.
- Define spatial properties of datasets.

Physical design of a database model
- Data field specification
- Implementation
- Populate the database

To undertake the database design steps in this section, we use a data modeling language called UML (Rumbaugh, Jacobson, and Booch 1999), and in particular the artifact called *class diagrams*, to create entity relationship models for the conceptual phase of database design. You have actually seen a conceptual class diagram previously in Figure 5.1. ESRI (2003) provides a UML tutorial showing how we can use the geodatabase data model to create a database model.

In the following subsections we develop a geodatabase design of the Green County project using nine steps categorized in terms of the three database model levels introduced previously. We provide an overview in Table 5.4 (adapted from Arctur and Zeiler 2004).

5.3.1 Conceptual Design of a Database Model

A conceptual data model language commonly comprises a simple set of symbols (e.g., rectangles for data constructs, lines for relationships, and bulleted labels for attributes) that can be used to compose a diagram to provide a means of communicating among database designers. We use the conceptual data model language to specify a conceptual database model, so we can understand the particulars of a database design domain, such as transportation planning or land use planning, or water resource planning. The simple diagrams are as close to everyday English as we can get without loading the diagrams with lots of implied, special meanings. We make use of the UML because ESRI has provided a utility to convert UML conceptual diagrams into logical schemas.

As part of the conceptual phase of database design, some of the steps use *data design patterns*, which are reoccurring relationships among data elements that appear so frequently that we tend to rely on their existence for interpretation of data. Data design patterns are similar to database abstractions identified 20 or so years ago (i.e., a relationship that is so important we commonly give it a label to provide a general meaning for the pattern). By the early 1980s, four database abstractions were identified in the semantic database management literature—classification, generalization, association, and aggregation (Nyerges 1991b). These four data abstractions relate directly to data design patterns and are used in the ArcGIS software (ArcGIS names are in parentheses): classification (Classification), association (Relationships), aggregation (Topology Dataset, Network Dataset, Survey Dataset, Raster Dataset), and generalization–specialization (Subtype). Such design patterns (abstractions) specify behaviors of objects within data classes to assist with information creation.

The products of a conceptual phase of database design help analysts and stakeholders discuss the intent and meaning of the data needed to derive information, placing that information in the context of evidence and knowledge creation; that is, both groups want to get it "right" as early as possible in the project, before too much energy is expended "down the wrong path."

1. *Identify the information products or the research question to be addressed.* Most every project has a purpose and requires a set of information products to address that purpose. To develop the best information available, identify the information products to be produced with the application(s). For example, a product might be a water resource, transportation, and/or land use *plan* as an array of community improvement projects over the next 20 years. Another could be a land development, water resource, or transportation improvement program that is a prioritized collection of projects within funding constraints over the next couple of years. The priority might simply be that we can fund only some of the projects among a total set of projects recommended for inclusion in an *improvement program*. A third product might be a report about social, economic, and/or environmental impacts expected as a result of the *implementation* of one or more of these projects in an improvement program.

A GIS data designer/analyst would converse with situation stakeholders rather than guess about the information outcomes to appear in the product. The stakeholders would then mull it over a bit to make sure they have an idea. Some guidance should be available in terms of a project statement. In the Green County project, the Council provided the purpose and the objectives in siting a wastewater treatment facility. In another context, perhaps the purpose is a research statement, in which one or more research questions have been posed. Sometimes such questions are called "need-to-know questions." For example, what do the stakeholders "need to know" about the geographical decision situation under investigation? What are the gaps in information, evidence, and/or knowledge? What information that is not available should be available to accomplish tasks related to decision situations? What changes (processes) in the world are important to the decision situation? What are the decision tasks? These questions should help the reader articulate "information needs" as a basis of data requirements.

To make this clearer, we use the Green County project, and discuss the values, goals, and objectives related to the criteria in Table 3.2. The overall task for the Green County GIS project is to find a suitable site for a wastewater treatment facility. What does *suitable* mean in terms of concerns about such facilities in the community—or, for that matter, any other public facility that results in community impacts? A fundamental question about this set of criteria is: Are those criteria sufficient to perform an appropriate site suitability study? If they are not, might there be unanswered community concerns? If so, might the wastewater siting project and eventual construction be stalled? How can we know this? We must start with the basics of data and understand how they relate to "community values"—not just to one or two groups, but to all groups that are impacted by the siting of a wastewater treatment facility.

Let us look at an example of a value structure that a citizens group might want to construct for a broad-based interpretation of the site suitability problem (Table 5.5). All cells in the table started out as someone's concerns. The upper half presents the values, goals, and objectives associated with the criteria from Chapter 3 (Table 3.2). The concerns were ordered in such a way as to make them practical relative to GIS-based decision problem solving. When a concern is labeled as an *objective* or *criterion*, it is then *measurable* as an attribute associated with spatial and temporal data to form the data elements of the GIS database.

Furthering this investigation about concerns, one can assume that clean water is an implicit concern, in that wastewater treatment plants are meant to clean water. We add a few more explicit concerns (values, goals, objectives, and criteria) at the bottom of Table 5.5 to flesh out some of the other issues that would make sense in this site selection. Clean water should certainly be an *explicit* concern. We know the site will tap into a sewer main at a junction point, but what about the treated water (i.e., the gray water)? Where is it to go? Commonly, a corridor is needed to route an outfall pipe. Because the treatment plant is to be outside the floodplain, but within 1,000 meters of the river, there must a pumping corridor that takes into consideration the impacts on parcels within and nearby the corridor. In addition, perhaps we can make use of the gray water for large water users who do not need the level of *clean* that is expected of drinking water.

A value structure is the foundation of the data categories, and forms the foundation of the "representation model." Both the value structure and data categories are needed, because they provide meaningful content for use by decision makers, technical specialists, and stakeholder groups in addressing a complex decision problem.

Sometimes people are able to define classes of data by knowing the subject matter, and simply writing out the class name and attributes. However, some people prefer to do some empirical work first (i.e., create instances of those classes by writing them out in a word processor table, spreadsheet, or just a text document). Then, they generalize those instances to create the field names and an object class specification. Once several feature classes have been entered, we can discuss the specification, including the relationships among those classes, which leads us to the next step.

2. *Identify the key thematic layers and feature classes.* A thematic layer is a superclass of information that commonly comprises a dataset(s) and perhaps several feature classes (hence, feature layers) convenient for human conversation about geographic data. For each thematic layer, specify the feature classes that compose that thematic layer. For each feature class, specify the potentially available data sources, spatial representation of the class, accuracy, symbolization, and annotation to satisfy the modeling, query,

TABLE 5.5. Simplified Value Structure for a Green County Database

Values	Goals/target	Objectives	Criteria for site suitability from Table 3.2
Community financial frugality is important.	Less than 365 meters	Minimizing the elevation leads to lower pumping costs.	Elevation measured in meters.
Community environmental health is important.	Outside the 100-year floodplain	The farther outside (up to a limit) the floodplain, the less susceptibility to ecosystem degradation.	Floodplain measured as elevation below mean sea level.
Financial frugality is important.	Lowest engineering costs possible, translated as 1,000 meters of river.	The closer to the river, the less cost to build outflow piping.	Within 1,000-meter distance of river.
Community environmental health is important.	Human health is protected when site is at least 150 meters from property.	The farther from residential property, the healthier the residents.	Euclidean distance in meters from residential property.
Green County landscape aesthetics are important.	At least 150 meters from parks.	The farther from park property, the better preservation of park aesthetics.	Euclidean distance in meters from parks.
Green County has some available property to serve as a location.	On vacant land.	Identify all vacant land parcels as an initial qualifier.	Land vacancy.
Financial frugality is important.	Within 1,000 meters of the wastewater pipe junction.	The closer to an existing wastewater pipe of appropriate diameter, the better.	Euclidean distance to wastewater pipe junction in meters.
Financial frugality is important.	Within 50 meters of a road.	The closer to an existing roadway, the better.	Euclidean distance to roads in meters.
Provide for needs of future residents, businesses, and industry.	At least 150,000 square meters.	Only those vacant land parcels that are large enough will do.	Square meters of parcels.

Additional concerns

Values	Goals/target	Objectives	Criteria for site suitability from Table 3.2
Clean water is needed for a healthy Green County community.	Locate site within 1,000 meters of large gray water users.	Maximize the availability of gray water.	Euclidean distances to large gray water users in meters.
Appropriate river flow makes for a healthy river.	River flows within 70–90 thousand cubic feet per second range are most preferred.	Minimum and maximum river flow should not fall outside 20,000–110,000 cubic feet per second.	River flows cubic feet per second.
Healthy fish communities help form a healthy human community.	Minimize the impacts on steelhead trout; no more than 1 trout in 1,000 to be impacted.	Maximize the improvement of water quality for all fish.	No more than X steelhead trout in X sections to be impacted.

and/or map product applications. For example, we might enumerate the information categories and data layers, as in Table 5.6. The data layers in the basic Green County GIS project are indicated in the second column with an asterisk (*), whereas the additional data layers needed for the enhanced analysis are indicated with a plus (+). The "information needs" in the first column drive the need for data.

3. *Detail all feature class(es)*. For each feature class, describe the spatial, attribute, and temporal data field names. For each feature class, specify the range of map scale for spatial representation and, hence, the associated spatial data object types. This determines whether multiple-resolution datasets for layers are needed. An analyst would have enough experience to know what resolutions of feature categories are appropriate for the substantive topic at hand. Revisit step 2 as needed to complete this specification. Identify the relationships among the feature classes.

A GIS database design analyst need not use UML to explore and compose the detail. However, this detail is documented in step 4.

At this point it is important to specify the spatial reference systems for the GIS database model as related to the geographic scale of the database. The decision focuses on geodetic, cadastral, and land survey information as related to coordinate systems. Different spatial reference systems use different coordinate systems based on how large a geographic area is required in the base map, and how accurate the coordinates must be for the application. Longitude and latitude coordinates that span the entire earth surface are often not suitable for urban and regional database model applications. Universal Transverse Mercator coordinate regions often cover larger areas than State Plane coordinate systems; thus, the accuracy of state plane coordinates are better.

4. *Group representations into datasets*. A *feature dataset* is a group of organized feature classes, based on relationships identified among the feature classes that help in generating information needed by stakeholders. The dataset creates the instance of a thematic layer, or a portion of the thematic layer, in which the relationships among feature classes are critical for deriving information. Analysts name feature classes and feature datasets in a manner convenient to promote shared understanding between themselves and stakeholders. Feature datasets are used to group feature classes for which topologies or networks are designed or edited simultaneously.

A feature dataset is but one of several data design patterns provided in the geodatabase data model. A *data design pattern* is a frequently occurring set of relationships that a software designer has decided to implement in a software system. Discrete features are modeled with feature datasets that comprise feature classes, but relationship classes, rules, and domains are three other design patterns. Continuous features are modeled with raster datasets. Measurement data are modeled with survey datasets. Surface data are modeled with raster and feature datasets. These other design patterns are used in more detailed database designs below.

Each feature dataset in the Green County conceptual overview makes use of a more detailed page diagram (Figure 5.4) to document the details of feature datasets in step 3.

TABLE 5.6. Geographic Information Categories and Data Layer Needs for the Original Green County GIS Project and an Enhanced GIS Project

Geographic information needs (based in part on Table 3.2) [original Green County Project + enhanced Green County Project]	Geographic data layer (based in part on Table 3.2) [original Green County GIS Project + enhanced Green County GIS Project]
* Basemap	* Coordinate reference system
Environmental characteristics + Soil characteristics	+ Layer: Soil series Source: NRCS Map use: display and analysis of soil characteristics
* Topography	* Layer: Elevation (DEM) Source: Green County DOT; USGS Map use: display and analysis of topographical terrain
* Water courses	* Layer: National Hydrographic Dataset Source: Green County; USGS and EPA Map use: display and analysis of surface water flows and water quality

* Land cover/use	Layers:
	Parcel boundaries
	    * Land use
	    + Site address
	Source: Green County
	
	+ Vegetation/land cover
	Source: Multiple agencies (EPA, USGS, BLM, various state agencies)
	Map use: display and analysis of vegetation land cover

+ Natural hazards	Layers: + Geohazards + Floodplain areas + Tsunami-prone areas + Historic wildfires Source: multiple agencies including USGS, U.S. Forest Service, state agencies Map use: display and analysis of natural hazard risk
* Ecologically sensitive areas	Layers: * Wetlands/lowlands * Parks Source: Green County + Protected areas + Protected habitats Sources: multiple agencies including USGS, EPA, U.S. Forest Services, Gap Analysis Program, state agencies Map use: display and analysis

(cont.)

TABLE 5.6. *(cont.)*

Geographic information needs (based in part on Table 3.2) [original Green County Project + enhanced Green County Project]	Geographic data layer (based in part on Table 3.2) [original Green County GIS Project + enhanced Green County GIS Project]
Infrastructure characteristics + Buildings * Transportation * Utilities	Layers: + Building footprints * Roads * Streets * Sewer lines Source: Green County
+ Other utilities	+ Gas and electricity lines + Water lines + Groundwater wells + Septic tanks + Landfills Source: various local land use and planning management agencies, DOT, local utility providers Map use: display and analysis
Land designations + Zoning	+ Layers: Restrictions on uses Source: local land use planning and management agencies Map use: display and analysis
Land administration + Boundaries	Layers: + Administrative areas + Cadastral framework (Public Lands Survey System) Sources: U.S. Bureau of the Census, local and regional land surveys Map use: display and analysis
Land ownership + Boundaries	Layers: + Ownership and taxation + Parcel boundaries + Survey network Source: local land use planning and management agencies, local land surveys Map use: display and analysis

Note. * in basic database design; + in enhanced database design; BLM, Bureau of Land Management; DEM, digital elevation model; DOT, Department of Transportation; EPA, Environmental Protection Agency; NRCS, Natural Resources Conservation Service; USGS, U.S. Geological Survey.

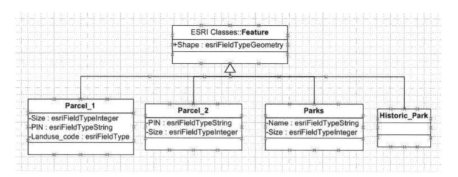

FIGURE 5.4. Land feature dataset package in the Green County conceptual database design.

5.3.2 Logical Design of a Database Model

Data processing operations to be performed on the spatial, attribute, and temporal data types individually or collectively derive the information (from data) to satisfy step 1. Such operations clarify the needs of the logical design.

  5. *Define attribute database structure and behavior for feature classes.* Apply subtypes to control behavior, create relationships with rules for association, and classifications for complex code domains.

• *Subtypes.* Subtypes of feature classes and tables preserve coarse-grained classes in a data model, and improve display performance, geoprocessing, and data management, while allowing a rich set of behaviors for features and objects. Subtypes let an analyst apply a classification system within a feature class and apply behavior through rules. Subtypes help to reduce the number of feature classes by consolidating descriptions among groups, improving performance of the database.

• *Relationships.* If the spatial and topological relationships are not quite suitable, then a general association relationship might be useful to relate features. Relationships can be used for referential integrity persistence, for improving performance of on-the-fly relates for editing, and for use with joins for labeling and symbolization.

  6. *Define spatial properties of datasets.* Specify rules to compose topology that enforces spatial integrity and shared geometry, and networks for connected systems of features. Topological and network rules are set to operate on features and objects. Set the spatial reference system for the dataset. Specify the survey datasets, if needed. Specify the raster datasets as appropriate.

• *Topology.* Topological rules are part of the geodatabase schema and work with a set of topological editing tools that enforce the rules. A feature class can participate in no more than one topology or network. Geodatabase topologies provide a rich set of configurable topology rules. Map topology makes it easy to edit the shared edges of feature geometries.

- *Networks.* Geometric networks offer quick tracing in network models. These rules that establish connections among feature types on a geometric level are different than the topological connectivity. Such rules establish how many edge connections at a junction are valid.

- *Survey data.* Survey datasets allow an analyst to integrate a survey control (computational) network with feature types to maintain rigor in the survey control network.

- *Raster data.* Analysts can introduce high-performance raster processing through raster design patterns. Raster design patterns allow them to aggregate rasters into one overall file, or maintain them separately.

5.3.3 Physical Design of a Database Model

7. *Data field specification.* For data fields, specify valid values and ranges for all domains, including feature code domains. Specify primary keys and types of indices.

- *Classifications and domains.* Simple classification systems can be implemented with coded value domains. However, an analyst can address complex (hierarchical) coding systems using valid value tables for further data integrity and editing support (Figure 5.5).

At this time, primary and secondary keys for the data fields are specified, based on valid domains of each field. A data key reduces the need to perform a "global search" on data elements in a data file. Hence, a key provides fast access to data records. A primary (data) key is used to provide access within the collection of features that can be distinguished by a unique identifier (Figure 5.6). When an analyst uses a primary key, he or she can easily distinguish one data record from another. A parcel identification code is

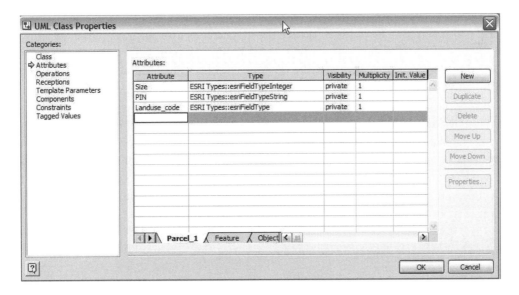

FIGURE 5.5. Data type specifications for land parcel data depicted with MS Office Visio UML class properties.

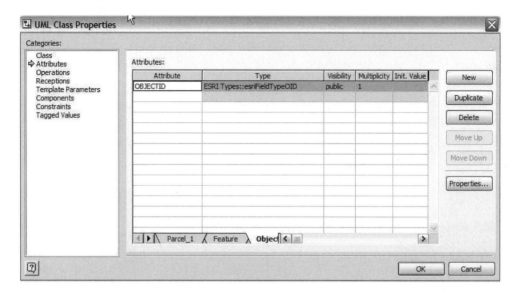

FIGURE 5.6. ObjectID for the primary key depicted with MS Office Visio UML class properties.

an example of a potential primary key for land parcel data records. A secondary key is used for data access when the data elements are not unique but are still useful to distinguish data records (e.g., land use codes). All land parcel data records of a particular land use code can be readily accessed.

8. *Implementation of schema.* Load the data schema into a database management system. Once loaded, a schema semantics check can be performed to ensure a computable schema. After running the semantics check and identifying errors, data schema modifications need to be made to rectify the described errors, and the semantics check should be run again. A database analyst would address as many errors as possible before rerunning the semantics check. If the error check results in no errors, or at least only errors that are tolerable, then an analyst can test the computability of the data schema.

9. *Populate the database*—not really part of design but a major part of development. The last step is to load data into the schema, often called *populating the database*. Coordinates underpin land records, transportation records, and water resource records in database models. Thus, the first step in implementing the GIS database model is to implement the coordinate reference system, although the reference system can be changed at any point throughout the project as appropriate.

5.4 Summary

Data modeling is a fundamental concern in GIS databases. It deals with data classes, information categories, evidence corroboration, and knowledge-building perspectives.

The purpose of elucidating these four levels of knowing is to provide readers a perspective that GIS is not just about data and databases, but that it extends through higher levels of knowing. Understanding these terms sets the stage for understanding data models and database models.

Data modeling is a process of creating database designs. A database design process creates several levels of database descriptions; some are oriented toward human communication, whereas others are oriented toward computer-based computation. The terms *conceptual*, *logical*, and *physical* have been used to differentiate levels of data modeling abstraction. A data model is the foundation framework that underlies the expression of a database model (i.e., we use data models to design database models). A database model is a particular design of a database (i.e., the design has some real-world substantive focus, not just an abstract expression of data constructs). A conceptual data model organizes and communicates the meaning of data categories in terms of object (entity) classes, attributes, and (potential) relationships. Logical data models (e.g., object, relational, or object–relational) are the underlying formal frames for database management system software. A physical data model expresses physical storage units and includes capabilities to specify performance enhancements, such as indexing mechanisms that sort data records. Each of these data models can have a corresponding database model for a particular set of information categories.

There are three special aspects to data models: constructs, operations (that establish relationships), and integrity/validity constraints (rules). Differences in data models dictate differences in data constructs used to store data, differences in operations on those data for retrieving and storing, plus differences in validity constraints used to ensure a robust database. ArcGIS software includes a large set (but still not all) data models: raster or image/grid data model, TIN data model, shapefile data model, coverage data model, and the geodatabase data model. The TIN and the grid are often used to represent continuous surfaces. The shapefile, coverage, and geodatabase data models are used for storing points, lines, and areas that represent mostly discrete features.

We use data models to create database models. Database models are the outcome of a database design process. We introduced a geodatabase database design process as a data modeling process comprising nine steps spread across the three levels—conceptual, logical, and physical—of data models. The conceptual design process that forms a conceptual database model comprises four steps: (1) identifying the information products or the research question to be addressed, (2) identifying the key thematic layers and feature classes, (3) detailing the feature class(es), and (4) grouping representations into datasets. The logical design process that forms a logical database model comprises two steps: (1) defining attribute database structure and behavior for feature classes, and (2) defining spatial properties of datasets. The physical design process that forms a physical database models comprises three steps: (1) data field specification, (2) implementation of the schema, and (3) populating the database. The outcome of that process was an extended Green County database design and database.

5.5 Review Questions

1. Differentiate among data, information, and evidence.

2. Why is it important to differentiate between evidence and knowledge?

3. Why is it useful to understand the difference between a data model and a database model when choosing a software system versus the data categories to develop an application?

4. Why do we have three levels of database abstraction—conceptual, logical, physical—models?

5. What are the three components of every conceptual, logical, and physical data model?

6. What is the difference between an image data model and grid data model?

7. Why did ESRI develop the geodatabase data model?

8. What is a general process for undertaking database design?

9. Why is a concerns hierarchy important to database design?

10. Describe the nature of a database design process in terms of the conceptual, logical, and physical database models that result from the process.

CHAPTER 6

Fundamentals of GIS-Based Data Analysis for Decision Support

In Chapter 3 we presented a workflow framework using an example of a decision situation involving a major investment decision process in which the goal was to find a suitable location for a wastewater facility within the Green County. We followed up with a decision situation assessment in Chapter 4, then data models in Chapter 5 that direct the possibilities for database design. That activity set the stage for the fundamentals of GIS-based data analysis for decision support presented in this chapter. More advanced work with multicriteria data analysis is presented in Chapter 7.

GIS-based data analysis is the core of "decision situation analysis" for each of the separate planning, improvement programming, and project implementation decision situations. Therefore, we focus on GIS data analysis from the perspective of how can we make appropriate choices for data analysis. Regardless of the complex relationships that might occur within and among decision situations, we can say that, fundamentally, the choice of GIS data analysis techniques, together with a decision analysis perspective for support of planning, improvement programming, and implementation analysis, depends ultimately on information needs.

In section 6.1 we discuss the direct connections between information and data analysis. In section 6.2 we present a framework for classifying GIS data analysis operations. The framework is based on (1) the geometry of spatial data, and (2) the spatial relationships used in GIS data analysis. The framework is complementary with other classifications, for example, the functional classification used in ArcGIS toolboxes, and it can be used as a guideline in choosing a GIS data analysis tool for a specific task. We use the framework in section 6.3, in which we present a workflow for finding suitable locations for Green County wastewater treatment facility. In the workflow, we link information needs and data requirements with GIS analytical operations to show how basic GIS data analysis can be used to address decision problems, but there are yet more advanced and interactive ways to treat the data.

6.1 Information Needs Motivate Data Development and Data Operations

Which information is needed for planning, programming, and implementation decision support depends on the content scoping of a decision-making process, as outlined in the decision assessment framework presented in Chapter 4. However, it is appropriate at this time to emphasize the bridge between information needs and data analysis. One can always use the data one has "on hand" to address a decision problem. It is also important to recognize that one's analysis might come up short in regard to generating information, and a decision must be made to pursue more data or simply recognize the uncertainty that is created when better data are not available.

Informational needs of plans, programs, and individual projects are addressed by transforming data into information through the use of data processing techniques. In this chapter we focus on the combination of data processing, using spatial analysis techniques implemented in GIS software. Much of the information needed to support decision-making activities requires geographical data about locations and their physical, social, economic, and environmental characteristics. Decision support can be conceptualized (albeit very simply) as a process of responding to information needs of decision activities by transforming data into information. The relationship presented in Figure 6.1, in which information needs guide the selection of geographical data and in turn the data transformation operations, is the basis of a data analysis framework for decision support. The purpose of the framework is to support planning, programming, and project implementation by matching information needs that are common to decision making.

Information need is identified early in the GIS workflow process of Chapter 3 (e.g., in section 3.2.1.1 Identify Project Objectives); hence, is a core issue in the decision situation assessment. It follows that the database design (data modeling) process establishes the data need to be used in GIS data analysis (i.e., the representation model). However, if this were all that is needed, then GIS mapping would take care of all decision support needs. Drawing on relationships derived from data is really what GIS data analysis is about; the result is information about relationships. Deriving information (i.e., establishing the character of the actual relationships among data elements) can be done through

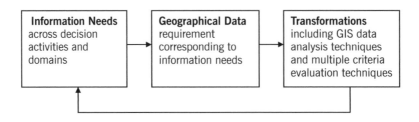

FIGURE 6.1. Making choices about data analysis relies on a relationship among information needs, geographical databases, and data transformations.

data transformations, which are considered the core of data analysis. Spatial analysis of geographical data involves many operations resulting in qualitative (categorical) and quantitative characteristics based on position, distance, direction, geographical extent, and shape. These spatial characteristics can be combined with nonspatial characteristics of locations (i.e., attributes) into effectiveness measures using multicriteria evaluation (MCE) techniques. Such effectiveness measures can become the basis for choosing among alternative planning scenarios in a planning decision situation, among funding scenarios to compose a budgeted program in improvement programming, and among alternative project versions to satisfy needs for an implementation decision situation.

6.2 Framework for Organizing and Selecting GIS Data Analysis Operations

It is important to recognize at the onset of our discussion that not all information needs identified during planning, improvement programming, and project implementation can be answered with GIS. Some information requires other data processing tools, such as spreadsheets, statistical software, or specialized models. For example, the calculation of project cost may require financial formulas that are available in spreadsheets. Similarly, the calculation of impacts on water quantity availability and water quality across various land use projects requires water flow and water quality simulation models. In regard to those information needs that involve spatial relationships and can be answered with GIS, it is not always clear which specific GIS data analysis operation, or chain of operations, one should select to arrive at the desired results. In the framework we present in this section, we aim to make the transition from formulating information needs to selecting appropriate GIS data analysis operations that are more transparent. The idea of a framework for organizing analytical GIS operations is not new. Various frameworks have been proposed that relate properties of spatial data types to analytical operations of GIS. Albrecht (1999) argued for a minimal and sufficient set of universal analytical GIS operations motivated by the difficulty posed by GIS software to all but sophisticated users. His minimal set includes 20 task-oriented analytical operations, including search, locational analysis, terrain analysis, distribution/neighborhood, and spatial analysis operations sufficient to solve many common analytical tasks. The missing link in Albrecht's framework is the lack of conceptual bridge to transition from users' information needs to analytical operations. Chrisman (2002) has offered a more comprehensive framework based on measurement frameworks linking space, time, and attributes; discovery of relationships among the data; and transformations of relationships. In his framework, the levels of data measurement (nominal, ordinal, interval, ratio, count, and absolute) are associated with generic types of analytical operations, such as rank, evaluate, rescale, cross-tabulate, classify, group, and so forth, which have their counterparts in specific GIS operations. Because data measurements can address users' information needs, Chrisman's framework does provide a transition from information needs to analytical operations that can supply needed information. Tobler (1979) presented a transformational view of cartography, and Chrisman (1999b) provided a transformational view of GIS operations, which we extend by cross-tabulating possible transformations among four common spatial data types: points, lines, polygons, and fields (see Table 6.1).

In Table 6.1, we list all 16 possible combinations of transformations among four spatial data types deriving from vector (points, lines, polygons) and raster (fields) data models. It is important to point out that transformations in Table 6.1 have a broader meaning than merely a change from one data type into another, because they also include interactions between data types, for example, a ground water well, represented by a point object interacting with an adjacent stream represented by a line object. Although the transformations among the spatial data types in Table 6.1 are straightforward, it is still difficult to relate them to user information needs.

We build on Tobler's and Chrisman's frameworks, and explicitly introduce spatial relationships to bridge the gap between information needs and data measurement frameworks. Spatial relationships derive from fundamental topological relations between spatial objects represented as empty or nonempty intersections of boundaries and interiors. Egenhofer and Franzosa (1991) identified nine such outcomes (relations) for two spatial regions that can be disjoint, touching, equal, containing, contained by, covering, covered by, overlapping with disjoint boundaries, and overlapping with intersecting boundaries. Relationships between spatial data derive from simple properties of distance, direction, and pattern, which are fundamental to spatial data analysis. They articulate common situations of real-world objects being positioned vis-à-vis each other at a certain distance, along a certain direction, and representing a certain arrangement that may be organized or disorganized. Many of these arrangements provide clues or outright answers to questions resulting from information needs. For example, the question of whether a specific project location encroaches on the area designated as the habitat of a protected species can be answered by evaluating whether both areas—the project footprint area and the habitat area—overlap. Overlap is just one specific spatial relationship. We list other spatial relationships below as a basic set of relationships that are important for many GIS operations, and provide definitions of these relationships from the *American Heritage Dictionary of the English Language* (2006):

- Connectedness—having a continuous path between any two points (nodes)
- Adjacency—contiguity or next to

TABLE 6.1. Spatial Data Transformations by Geospatial Data Types

From:	To:			
	Points	Lines	Polygons	Fields
Points	Points → points	Points → lines	Points → polygons	Points → fields
Lines	Lines → points	Lines → lines	Lines → polygons	Lines → fields
Polygons	Polygons → points	Polygons → lines	Polygons → polygons	Polygons → fields
Fields	Fields → points	Fields → lines	Fields → polygons	Fields → fields

- Containment—inside or within
- Proximity—nearness, closeness
- Overlap—lay over and/or covering
- Pattern—a sequence of systematic spatial occurrences in at least two dimensions
- Flow—movement through/across space and/or through a channel

We can now combine GIS transformations from Table 6.1 with spatial relationships, thus addressing information needs that arise from specific decision situations to be addressed by GIS data analysis. Using the example of wastewater facility project locations and protected species habitat we can address the question of which project locations do not encroach on protected species habitat by combining the overlap relationship with polygon-on-polygon transformation. The GIS data analysis operation that finds polygon-on-polygon overlaps is polygon-on-polygon *overlay intersection*. Notice that we still need "mentally" to relate the specific information need—"find project locations that do not encroach onto the protected species habitat"—to an appropriate combination of spatial data transformation and relationship. However, once we can link the information need with the appropriate combination, the challenge of finding a suitable GIS data analysis operation(s) becomes easier to overcome. To help us in the task of finding the appropriate GIS data analysis operations we enumerate all combinations of spatial relationships with spatial data transformations. They are listed by spatial data types (points, lines, polygons, and fields) in Tables 6.2 through 6.5. GIS data analysis operations addressing the combinations are given in table cells. We use in the tables the operation names implemented in ArcGIS 9.x software. An empty cell means that a GIS data analysis tool does not exist for the given combination of spatial relationship and transformation. The fact that there are empty cells in the tables underscores the need for abstraction and generalization of spatial data, and "analysis chains." For example, it may not be possible to compute directly spatial relationships for line-to-point transformations, but it is possible to represent (abstract) a line by a series of points, thus representing the line-to-point relationships on the point-to-point bases, and thereby use multiple operations in sequence. These multiple operations in sequence constitute a finer grain of workflow within each of the modeling phases presented in Chapter 3, and at the same time provide an ability to create other types of models—for example, as constituted in ESRI's ModelBuilder workflow environment.

The reader may notice the conspicuous absence of GIS operations in many of the cells in Table 6.5. These cells correspond to field → vector transformations combined with spatial relationships. Even though GIS software provides ample tools for raster-to-vector transformations, which amounts to creating discrete objects from field representations of continuous phenomena, only a few tools in modern GIS address spatial relationships in such transformations. This is perhaps emblematic of what Goodchild, Yuan, and Cova (2007) call "the most problematic area of geographic data modeling in current GIS practice" (p. 257)—the process of discretization of phenomena that are essentially continuous on the Earth's surface.

TABLE 6.2. GIS Data Analysis Operations for Computing Point-Based Spatial Relationships

Relationships	Transformations			
	Points → points	Points → lines	Points → polygons	Points → fields
Connectedness		Locate features along routes		Spatial interpolation
	Select by location: point features that *intersect* with point features	Select by location: point features that *intersect* with line features	Select by location: point features that *intersect* with polygon features	
Adjacency				Spatial interpolation
Containment		Overlay intersect point-on-line	Overlay intersect point-on-polygon	Neighborhood point statistics
	Select by location: point features that *have their center in* OR *are contained by* point features	Select by location: point features that *have their center in* OR *are contained by* line features	Select by location: point features that *have their center in* OR *are contained by* OR *are completely within* polygon features	
Proximity	Point distance, near, buffer	Near, buffer	Buffer	Map algebra neighborhood operations
	Select by location: point features that *are within a distance* of point features	Select by location: point features that *are within a distance* of line features	Select by location: point features that *are within a distance* of polygon features	
Overlap	Merge	Overlay intersect point-on-line,		Map algebra zonal operations
	Select by location: point features that *are identical to* point features	Locate features along routes		
Pattern	Point pattern spatial statistics: Average nearest neighbor, Moran's index, Getis–Ord general G tool			Point density, kernel density, spatial interpolation
Flow		Network analyst: New route, create network location, solve (shortest path)		Path distance, Path allocation

TABLE 6.3. GIS Data Analysis Operations for Computing Line-Based Spatial Relationships

Relationships	Transformations			
	Lines → points	Lines → lines	Lines → polygons	Lines → fields
Connectedness	Select by location: line features that *intersect* with point features	Overlay intersect line-on-line Select by location: line features that *intersect* with line features	Overlay intersect line-on-polygon Select by location: line features that *intersect* with polygon features	Euclidean distance, path distance
Adjacency		Select by location: line features that *share a line segment* OR *touch the boundary* with line features	Select by location: line features that *share a line segment* OR *touch the boundary* with polygon features	
Containment		Select by location: line features that *have their center in* OR *are contained by* line features	Overlay intersect line-on-polygon Select by location: line features that *have their center in* OR *are contained by* OR *are completely within* polygon features	
Proximity	Buffer Select by location: line features that *are within a distance* of point features	Buffer Select by location: line features that *are within a distance* of line features	Buffer Select by location: line features that *are within a distance* of polygon features	
Overlap		Merge Overlay intersect line-on-line Locate features along routes Select by location: line features that *are identical to* line features		
Pattern		Directional mean, Directional distribution		Neighborhood line statistics
Flow		Network analysis: New route, Create network location, solve (shortest path)		

TABLE 6.4. GIS Data Analysis Operations for Computing Polygon-Based Spatial Relationships

| Relationships | Transformations | | | |
	Polygons → points	Polygons → lines	Polygons → polygons	Polygons → fields
Connectedness	Select by location: polygon features that *intersect* with point features	Select by location: polygon features that *intersect* with line features	Overlay intersect polygon-on-polygon Select by location: polygon features that *intersect* with polygon features	
Adjacency		Select by location: polygon features that *share a line segment* OR *touch the boundary* with line features	Merge Select by location: polygon features that *share a line segment* OR *touch the boundary* with polygon features OR *dissolve a common boundary*	
Containment			Clip, select Select by location: polygon features that *have their center in* OR *contain* OR *completely contain* OR *are contained by* OR *are completely within* polygon features	Raster clip
Proximity	Buffer Select by location: polygon features that *are within a distance* of point features	Buffer Select by location: polygon features that *are within a distance* of line features	Buffer Select by location: polygon features that *are within a distance* of polygon features	
Overlap			Polygon overlay: intersect, union, identity, update merge Select by location: polygon features that *are identical to* polygon features	
Pattern			Spatial statistics: Moran's index General G	
Flow				

TABLE 6.5. GIS Data Analysis Operations for Computing Field-Based Spatial Relationships

Relationships	Fields → points	Fields → lines	Fields → polygons	Fields → fields
		Transformations		
Connectedness				Map algebra, distance, zonal operations
Adjacency				Map algebra, neighborhood operations
Containment	Extract (cells) by points		Extract (cells) by circle, polygon	Map algebra, zonal, neighborhood operations
Proximity				Map algebra, distance operations
Overlap	Extract (cells) by points		Extract (cells) by circle, polygon	Map algebra, local, multivariate operations
Pattern				Map algebra
Flow				Flow direction, Flow accumulation

Some user data needs may require spatial transformations before one can combine them with spatial relationships for the simple reason that no GIS data analysis operations address a given combination of data transformation and relationship. Take, for example, the combination "Points → fields" and "Proximity" in Table 6.2. Such a combination may be useful for computing the distance between watershed and the nearest rainfall gauge measuring the amount of precipitation. Unfortunately, there is no specific GIS data analysis operation for computing the distance between a point location and a zone of raster cells. We can, however, transform zonal fields to polygons, then represent polygons by their centroids. We can then use the "Points→points" and "Proximity" combination and use the *point distance* operation (Table 6.2). To represent polygons as their centroids we should transform polygons to points. This transformation, along with nine other vector-to-vector feature transformations listed in Table 6.1, can be accomplished in ArcGIS using tools from *Features* toolset located in the *Data Management Tools* toolbox. The field-to-polygon transformations in Table 6.1 can be handled by transforming field (raster) data to polygon feature class using *From Raster* toolset in the *Conversion Tools* toolbox.

The framework presented in Tables 6.2 through 6.5 can be used to help select appropriate GIS data analysis tools to satisfy user information needs. In section 6.5 of this chapter we demonstrate the use of the framework on the example of finding and evaluating suitable locations for a wastewater facility in Green County. The GIS data analysis tools listed in Tables 6.2 through 6.5 are helpful in finding suitable locations. Before that, however, we present in the next section the data analysis workflow for Green County to find and evaluate suitable locations for a wastewater facility.

6.3 Information Needs, Data Requirements, and Data Operations in Green County Data Analysis

Back in Chapter 3 we indicated that the Green County GIS project is about area planning. In the following section we present the analysis of the Green County site selection project workflow and compare–contrast it with a more nuanced analysis workflow using several tables, one for each of the six phases of the landscape analysis workflow process, and use the same information need, data, and data operations framework in Figure 6.1. In Chapter 3, section 3.2, we introduced several questions that can be posed to guide GIS work at each phase. We take advantage of those questions below when framing the information needs for each of the six phases of the landscape modeling process contained in the Steinitz Framework presented below. We lead off with the representation modeling workflow, even though this kind of work is called preanalysis or preparatory analysis in some workflows. We present it here for completeness. Subsections that follow provide a detailed workflow description for each of the six models in the Steinitz modeling framework:

- *Representation*—a description of the wastewater facility conditions in the world (local area or region)
- *Process*—how the wastewater world changes states of the conditions in the representation model
- *Scenario*—what aspects of the world to consider; what conditions to address in the process
- *Change*—a before and after description (state) given those conditions that have changed
- *Impact*—the results/implications of that change on other conditions create an impact
- *Decision*—what choice of impacts (good and bad) one is minimizing or maximizing

6.3.1 Representation Model Workflow

From Table 3.5 in Chapter 3, workflow for representation modeling involves at least three substeps—problem description, database design, and database implementation. The problem description substep, which would have been performed as part of data needs analysis, establishes the "common ground" for the project; hence, the objectives handed to GIS analysts of Green County play this role. The second substep is database design, and considerable effort goes into preparing a database schema and assembling data dictionaries. The third substep is what we are most concerned about here, having addressed the first two substeps in Chapter 3 sufficiently. The database implementation has a direct connection to data analysis; hence, the most fundamental question to guide this representation model: Do we have the data in the appropriate form, or are data conversions necessary? Comparing the Green County workflow data analysis with nuanced landscape modeling data analysis workflow (from Chapter 3) we get the designations indicated in column 2 (points) and 3 (lines) of Table 6.1. The table entries of Table 6.6 with an asterisk (*) prefix are part of a basic stakeholder data need; whereas the table entries with a plus (+) prefix indicate data that could be used for an enhanced analysis. Remember, textbooks show and demonstrate basic principles rather than full details of actual projects. The entries with + are our own recommendations for a realistic GIS data analysis workflow.

Development of a representation model for the Green County wastewater facility site selection problem involves a number of datasets and operations. The National Hydrography Dataset (NHD) website can provide river data (as mentioned in Table 6.6) if a local river dataset does not exist (U.S. Geological Survey 2006a). This dataset contains nationwide coverage for the conterminous United States. The *define* coordinate system operation mentioned in the context of river dataset in Table 6.6 (third column) provides a way to rectify coordinate information, such as transforming from longitude–latitude to state plane coordinates. Several *hydrology* GIS tools exist in ArcGIS's Hydrology Toolset to manipulate NHD shapefiles (U.S. Geological Survey 2006b).

The land parcel data layer in the wastewater facility application makes use of the *merge* operation, which brings two data layers together into a single layer, commonly edge to edge.

The soils data referenced in Table 6.6 can be accessed from the U.S. Department of Agriculture, National Resources Conservation Service (2006a). A portal serves up maps for all areas of the United States (National Resources Conservation Service 2006b).

6.3.2 Process Model Workflow

When developing a process model, a fundamental question would be: What are the relationships among the spatial–temporal elements so that we have a better understanding of urban–regional process (i.e., what [who] influences what [who]? Basic relationships among parcels and other features (river, floodplain, parks, etc.) are incorporated in the criteria within Table 3.2, as given to an analyst, and described in Table 6.7 in terms of data analysis steps. The relationships are stated in those criteria, but they are in binary exclusionary and inclusionary form. We have no gradation of influence. Proximity operations (buffer, near, and point distance) are the first approximation of a process. However, those operations do not provide a "matter of degree" influence, but they nonetheless allow one to work with relationships that identify "suitable parcels."

TABLE 6.6. Representation Model of Information Needs, Data Requirements, and Data Operations

Information need based in part on Table 3.2 [* basic stakeholder need + enhanced need]	Data requirement [* basic stakeholder data + enhanced data]	GIS operations
Environmental characteristics		
* Close to water course	Layer(s): * Rivers Source: Green County's State DOT + National Hydrography Dataset Source: USGS and EPA	* Define coordinate system * Project to city coordinate system * Export to geodatabase * Define coordinate system + Define coordinate system
* Topography—below 365 feet	Layer: * Elevation (DEM) Source: Green County DOT; USGS	* Define coordinate system
* Land cover/use	Layers: * Parcel boundaries * Land use Source: Green County + Vegetation/land cover Source: Multiple agencies (EPA, USGS, BLM, various state agencies)	* Merge tiles into geodatabase feature class
* /+ Natural hazards	Layers: + Geohazards * Floodplain areas + Tsunami-prone areas + Historic wildfires Source: multiple agencies, including USGS, U.S. Forest Service, state agencies	+ Identify appropriate coordinate system * None needed
* Ecologically sensitive areas	Layers: * Lowlands/wetlands * Parks Source: Green County + Protected areas + Protected habitats Sources: multiple agencies, including USGS, EPA, U.S. Forest Services, Gap Analysis Program, state agencies	* Define coordinate system * Update with new historic park
+ Soil characteristics	+ Layer: Soil series Source: NRCS	+ Select from soil series stable soils

(cont.)

TABLE 6.6. *(cont.)*

Information need based on part on Table 3.2 [* basic stakeholder need + enhanced need]	Data requirement [* basic stakeholder data + enhanced data]	GIS operations
Infrastructure characteristics		
+ Buildings	Layers:	None needed
* Transportation	+ Building footprints	None needed
* Utilities	* Roads	+ Identify appropriate
	+ Streets	coordinate system
	* Sewer lines	
	Source: Green County	
	+ Gas and electricity lines	+ Identify appropriate
	+ Water lines	coordinate system
	+ Ground water wells	
	+ Septic tanks	
	+ Landfills	
	Source: various local land use and planning management agencies, DOT, local utility providers	
	Map use: display and analysis	
Land designations		
+ Zoning	+ Layers: Restrictions on uses	+ Identify appropriate
	Source: local land use planning and management agencies	coordinate system
	Map use: display and analysis	
Land administration		
+ Boundaries	Layers:	+ Identify appropriate
	+ Administrative areas	coordinate system
	+ Cadastral framework (Public Lands Survey System)	
	Sources: U.S. Bureau of the Census local and regional land surveys	
	Map use: display and analysis	

The process model for the Green County wastewater suitability analysis involves a number of operations and datasets. Most of the operations are based on proximity relationship, including the proximity of rivers, the proximity of elevation zones, the proximity of lowlands and wetlands, the proximity of residential parcels and parks, and the proximity of floodplain. This kind of process analysis is based on meeting exclusionary or inclusionary criteria. Also, observe that this kind of process analysis is purely a structural process that does not contain temporality. Consequently, it is not a true process model.

TABLE 6.7. Process Model Information Needs, Data Requirements, and Data Operations

Information need based in part on Table 3.2 [* basic Green County need + enhanced Green County need]	Data requirement [* basic Green County data + enhanced Green County data]	GIS operations
Environmental characteristics		
* Water courses	Layer(s): * Rivers Source: Green County's State DOT	Overlay analysis * Proximity—exclude from Buffer: river
	+ National Hydrographic Dataset Source: USGS and EPA	+ Select critical water courses
+ Noise, smell	Layer: Location-based noise measurements at sample locations	+ Overlay analysis * Proximity—delineate inclusionary zones Buffer
* Topography	Layer: * Elevation (DEM) Source: Green County DOT; USGS	* Extract by attributes (from DEM-based grid) * Transform (extracted zones) from raster to polygon * Proximity operations Buffer (lowland polygon) * Overlay: Intersect river buffer with lowland
* Critical resource areas	Layers: * Lowlands/Wetlands * Parks Source: Green County	* Buffer wetlands and parks + Buffer protected areas and habitats
	+ Protected areas + Protected habitats Sources: multiple agencies including USGS, EPA, U.S. Forest Services, Gap Analysis Program, state agencies	
* /+ Land cover/use	Layers: * Parcel boundaries * Land use + Site address Source: Green County	Analysis-extract operations * Select residential parcels * Merge parcels with park Proximity operations * Buffer residential and park
* /+ Natural hazards	+ Vegetation/land cover Source: Multiple agencies (EPA, USGS, BLM, various state agencies) Layers: + Geohazards * Floodplain areas + Tsunami-prone areas + Historic wildfires Source: multiple agencies, including USGS, U.S. Forest Service, state agencies	Proximity * Buffer natural hazard areas to create exclusionary zones * Buffer vegetation/land use categories to create exclusionary zones
+ Soil characteristics	+ Layer: Soil series Source: NRCS	Overlay analysis + Intersect stable soils with buffered hazards and vegetation/land use categories

6.3.3 Scenario Model Workflow

When developing a scenario model, a fundamental question would be: How does one assess whether the current state of the urban–regional environment is working well? The basic GIS project workflow did not consider any metrics of assessment to understand the current and future condition of the community. Metrics might include water flow (cubic feet per second increase–decrease) in the river, acres of lost–gained habitat, and average number of workdays increase–decrease in sick days due to ingestion of contaminated water. To consider these issues a process model must be reasonably robust, or assumptions must be made with regard to increase–decrease of conditions. Information needs, data requirements, and the data operations for the scenario model appear in Table 6.8.

TABLE 6.8. Scenario Model Information Needs, Data Requirements, and Data Operations

Information need based in part on Table 3.2 [* basic Green County need + enhanced Green County need]	Data requirement [* basic Green County data + enhanced Green County data]	GIS operations
* Suitable parcels	Layers: * Parcel boundaries * Land use Source: Green County	Spatial queries: Select by location * Select parcels *completely within* critical buffer, then switch selection * Select parcels that *are within a distance* of: Roads at 50 meters Wastewater piping junction at 500 and 1,000 meters
+ Increase–decrease flow in water courses (cubic feet per second)	Layer(s): + National Hydrographic Dataset + DEM + Unified Rain Gauge Dataset Sources: USGS, EPA, National Oceanic and Atmospheric Administration (NOAA)	+ Hydrology toolset: Flow Accumulation tool + Rational method model implemented with map algebra
+ Increase–decrease in habitat (acres)	Layers: + Protected areas + Protected habitats Sources: multiple agencies including USGS, EPA, U.S. Forest Services, Gap Analysis Program, state agencies	+ Attribute table field statistics Frequency Summary statistics (mean, mode, median)
+ Increase–decrease average number of sick days	+ Layer representing expected number of sick days per each zone around the plant; based on the distribution of population near the plant and susceptibility rates to plant-caused health problems	+ Spatial interpolation

The purpose of this information is to ascertain the feasibility/plausibility of a scenario. This means that GIS operations are needed to compare datasets with attribute values comprising the scenarios. In Table 6.8 we show that spatial query operations are used to identify parcel boundaries suitable for the wastewater facility site. These parcels are identified based on satisfying proximity conditions, by being either over or under, or at the prescribed distance from other spatial entities, such as specific land use locations, rivers, or piping junctions.

One can create alternative scenarios for different stream flow conditions, protected habitat areas, and number of sick days experienced by the population living in proximity of the plant. The GIS operations are sufficient to compute a first-cut approximation of flow in streams by using map algebra operations. Specifically one can multiply flow accumulation surface by daily or monthly average precipitation surface obtained from a unified rain gauge dataset, thus obtaining a rough estimate of stream flow under different precipitation scenarios. A step toward obtaining a more realistic estimate of stream flow would be to implement in GIS a simple model that takes into account spatial variability of land cover in addition to precipitation surface. One such model, called a *rational method model*, calculates the peak daily surface runoff rate based on a simple equation:

$$Q = kCiA \tag{6.1}$$

where Q is the flow rate (cubic meters/day), k is the conversion factor (0.0254), C is a dimensionless runoff coefficient, i is rainfall intensity (inches/day), and A is the area of each cell (square meters). The rational method is generally considered to be an approximate model for computing the flood peak resulting from a given rainfall, with the runoff coefficient accounting for all differences between the rainfall intensity and the flood peak. Such differences result from infiltration, temporary storage, and other losses. The runoff coefficient values range from 0 to 1, with rough surface areas, such as cropland or pasture, having small coefficient values and impervious areas, such as mixed urban or commercial land, having high coefficient values. Notice that flow rate Q can be easily computed using map algebra and multiplying raster layers representing the right-hand side of Equation 6.1.

To extend this scenario analysis, we could use a combination of flooding model and queries to answer the following questions: is it feasible to have increased flows that will not result in flooding, or is there enough suitable land to increase the habitat? Then, we could use field statistics to provide additional information for the scenarios.

6.3.4 Change Model Workflow

At least two important types of change should be considered. One is how the landscape might be changed by current trends. Modeling trends leads us to develop projection (population, employment, etc.) models that act as the basis of landscape change, even within a small area of a community. Another is how a community might be changed by implementation of a design action. A *design action* is an intervention, an activity that proactively changes a place rather than accepting change as "inevitable." That is, the change is directed by design. The types of information we can add to the workflow process in relation to a change model are indicated in Table 6.9. It would be good to know employment and population growth to provide basic information about the real need of a wastewater facility in the future. Employment and population growth might result in changes in water quality. GIS operations and non-GIS statistical operations addressing changes are among the entries in the right-hand column of Table 6.9.

For a change model, we are looking for attribute data to compute wastewater production rate. This rate can be computed from a deterministic function used by civil engineers. The rates can be multiplied by assumed population growth using simple algebraic functions (addition, multiplication) and working with table attributes (Field Calculator in ArcMap) to compute the wastewater production volume. Computing water quality degradation is more difficult, because it involves spatial–temporal processes that are normally modeled by handling differential equations in specialized models outside GIS (e.g., QUAL2 and a number of water quality models).

6.3.5 Impact Model Workflow

A change model forms the basis of what to consider for an impact model. Some important questions for the impact model follow: What are the predictable impacts (i.e., the outcomes based on the changes)? What influence would a change in the population

TABLE 6.9. Change Model Information Needs, Data Requirements, and Data Operations

Information need based in part on Table 3.2 [* basic Green County need + enhanced Green County need]	Data requirement [* basic Green County data + enhanced Green County data]	GIS operations
+ Employment/population growth projections	Layers: + Administrative areas + Cadastral framework (Public Lands Survey System) Sources: U.S. Bureau of the Census local and regional land surveys	+ Extract Operations • Clip • Select • Split + Spatial and attribute query Selection: • By attribute • By location • By graphics
+ Wastewater production from community use (based on employment and population growth projections)	Layer: + Dataset based on engineering standards for water resources. Source: USGS and EPA	+ Deterministic formulas implemented in ArcMap's Field Calculator
+ Water quality degradation rate by section of stream	Layer: + Sample from field Source: USGS and EPA	+ Water quality models; output visualized in GIS

TABLE 6.10. Impact Model Information Needs, Data Requirements, and Data Operations

Information need based in part on Table 3.2 [* basic Green County need + enhanced Green County need]	Data requirement [* basic Green County data + enhanced Green County data]	GIS operations
+ Wastewater production increase over the current level (based on employment and population growth change)	Layer: + Increase–decrease of wastewater production per area unit Sources: engineering wastewater production tables, EPA	+ Map algebra operations + ArcMap's Field Calculator
+ Water quality degradation by section of stream	Layer: + Field sample Sources: USGS and EPA	+ Water quality models; output visualized in GIS

growth have on the production of wastewater and, hence, on the size of the plant and the size of the parcels needed to site the wastewater plant? GIS operations that provide such information are among the entries in the right-hand column of Table 6.10.

One needs to answer three types of questions to address impacts: what (quantity or quality impact level), where (impact location), and when (temporal distribution of impact)? The levels of wastewater production and water quality degradation should be computed as spatially dependent functions, where input variables, such as predicted population density, are spatially distributed. At the simplest level, one would use algebraic operations available in ArcMap's Field Calculator to compute the volume of wastewater per each enumeration map unit (e.g., block) to come up with the total per area. However, the wastewater does not travel over land or in streams but along the sewer network, at least up to a certain point. We use the sewer network map for the Green County area, because we do not assume that all of the wastewater travels across the stream network.

Water quality usually computed with the use of complex functions (often based on differential equations) that would be tough to implement in the Field Calculator. In summary, we can use simple algebraic operations and functions in Field Calculator to compute quantity of impact and its spatial distribution. Computations that require complex functions incorporating temporal processes (e.g., water quality) need to be calculated with specialized numerical models outside of GIS.

6.3.6 Decision Model Workflow

A decision to site a wastewater treatment plant is based on information about impacts, and perhaps the trade-offs among impacts. Impacts must be mitigated. Several issues about equity can be addressed (e.g., greatest good for greatest number of people, or greatest good for those who really need the service the most). These issues were not considered in the Green County project except by assumption. Thus, a fundamental question is: How are we to treat impacts in an equitable manner considering that there are different types of equity? The assumption is that everyone is impacted equally as long as we use a particular distance buffer to protect people and land—the exclusionary criteria. The information needs, data, and operations for the decision model are provided in Table 6.11.

Multiple solutions can result from implementing the previously discussed decision model in GIS. Just consider the outcomes of incorporating in the model data prefixed with the + sign in the first column of Table 6.11. One will very likely arrive at dif-

TABLE 6.11. Decision Model Information Needs, Data Requirements, and Data Operations

Information need based in part on Table 3.2 [* basic Green County need + enhanced Green County need]	Data requirement [* basic Green County data + enhanced Green County data]	GIS operations
* Highly suitable parcel	Layers: * Parcel boundaries * Vacant land use Source: Green County	* Select by location: Select the parcels within a distance of 50 meters of road and within a distance of 500 meters of junction + Select adjacent parcels with touching boundaries
* Parcel recommendation	* Parcel boundaries * Parcel size Source: Green County	* Attribute table field summary statistics: Aggregate areas of contiguous parcels
+ Increase–decrease flow in water courses (cubic feet per second)	Layer(s): + National Hydrographic Dataset Sources: USGS and EPA	+ Hydrology Toolset: Flow Accumulation tool + Rational method model
+ Increase–decrease in habitat (acres)	Layers: + Protected areas + Protected habitats Sources: multiple agencies, including USGS, EPA, U.S. Forest Services, Gap Analysis Program, state agencies	+ Statistics operations • Frequency • Regression + Statistics operations • Frequency • Regression
+ Increase–decrease average number of sick days	+ Layer representing expected number of sick days per each zone around the plant	+ Spatial interpolation
+ Wastewater production increase over what currently exists (based on employment and population growth change)	Layer: + Engineering standards for water resources Sources: USGS and EPA	+ Output from impact model
+ Water quality degradation by section of stream	Layer: + Field samples Sources: USGS and EPA	+ Output from impact model

ferent suitable parcel locations based on different alternative results for stream water flows, protected habitat areas, number of sick days, wastewater production, and water quality. Note that alternative results arise as the consequence of considering different future conditions in scenario, change, and impact models. Because the future is generally uncertain, it is a prudent approach to consider alternative futures as opposed to a single future. But how is one to discriminate between alternative decision model outcomes? Which is better, which is worse? This question may be hard to answer, because the alternative model outcomes may involve trade-offs (e.g., one outcome may offer a suitable parcel adjacent to a sewer pipe, whereas another may offer a larger-area, suitable parcel farther away from sewer pipes). Resolving such trade-offs requires a systematic evaluation of alternative model solutions, also called *decision options* or *decision alternatives*, on the bases of measurable attributes called *criteria* and standardized outcomes for each decision option. The methodology developed specifically for the purpose of evaluating decision options on the basis of evaluation criteria, called *multiple criteria evaluation* (Voogd, 1983), has become an important GIS capability, adopted in the 1990s for the purpose of evaluating spatial decision alternatives commonly generated in GIS (Carver 1991; Jankowski 1995; Malczewski 1999). We discuss multiple criteria evaluation in Chapter 7.

6.4 Summary

It seems straightforward to write about the three approaches to choice making in a spatial decision problem: the conventional GIS data analysis suitability approach and a six-phase landscape analysis approach, and the connections between them. However, because so many people, so many organizations, and so much money are involved within decision processes in the workday world, there is a challenge to develop an integrated decision environment. For this reason, GIS data analysis has enormous potential to integrate perspectives across decision situations, thus fostering a sustainability management approach to societal transformation. In this chapter, we started from a framework that can facilitate an integrated decision environment by linking information needs with data requirements, and GIS data analysis operations that in turn can answer the information needs. We then presented the workflow for a six-phase analysis approach aimed at producing richer and more comprehensive data about the spatial choice problem than a conventional GIS suitability analysis. A more advanced treatment of multicriteria evaluation analysis is presented in Chapter 7.

6.5 Review Questions

1. How does the analysis of user information needs relate to GIS operations?

2. How does combining spatial data transformations and spatial relationships lead to

a framework for linking user information needs with GIS operations to address these needs?

3. Using the example of siting the Green County wastewater treatment facility, what is the advantage of articulating a difference among the models described using the Steinitz modeling framework?

4. What is the principal focus of a representation model?

5. What is the principal focus of a system process model?

6. What is the principal focus of a scenario model?

7. What is the principal focus of a change model?

8. What is the principal focus of an impact model?

9. What is the principal focus of a decision model?

10. What happens with information uncertainty when rather than implement one or more of the models, we make assumptions about the information outcomes of that model?

CHAPTER 7

Making Choices
about GIS-Based Multicriteria Evaluation

In Chapter 6 we presented a basic, GIS-only approach to evaluating suitable locations. That approach established an opportunity to introduce multicriteria evaluation (MCE) techniques to supplement decision support capabilities in this chapter. Based on our earlier presentation of a framework for classifying GIS data analysis operations and workflow to find suitable locations, we now introduce MCE methods, techniques, and tools that can help us sort, rank, and select from different decision options.

The equations that underlie techniques for MCE are not commonly available in ArcGIS, with the exception of weighted summation-based overlay. MCE techniques are implemented in two GIS software packages, including IDRISI (Eastman, Jin, Kyem, and Toledano 1995) and CommonGIS (Andrienko, Andrienko, and Gatalsky 2000). The authors also implemented them in a spatial decision support software for group decision making called GeoChoicePerspectives (Jankowski and Nyerges 2001b). We provide the equations to offer additional insight into just how decision options are evaluated in MCE.

MCE involves transformations of data, which characterize impacts of decision alternatives, resulting in a *summary score*, also called a *final appraisal score*. The idea of computing a final appraisal score is to provide one measure used as the basis for rank ordering of decision alternatives from best to worst. Three types of data transformations are common to each MCE technique: standardization of data to a common scale, transformation of decision maker preferences into numeric weights, and aggregation of standardized data with numeric weights into a common measure of intrinsic value—a summary score. Here we present an overview of these transformations in detail.

7.1 Data Standardization in MCE

In MCE, the attributes that are characteristic of decision alternatives and have known preference order of attribute values are used as the bases for evaluating the alternatives. The *preference order* means that, for example, a high attribute value is deemed to be better than the medium attribute value, which is in turn considered still better than the low attribute value; or, conversely, as in the case of a cost-related attribute, a low attribute value is deemed to be better than medium and high attribute values. Attributes with known preference order are called *evaluation criteria*, shortened to *criteria* in MCE. Criteria can represent various measurement scales. This creates a problem of dealing with the proverbial apples and oranges measurement scale comparison when evaluating decision alternatives. Just think of evaluating alternative landfill sites on diverse criteria such as permeability, land cover class, and distance to the nearest paved road. Because each criterion uses a different scale of measurement, criterion values cannot be easily aggregated into one overall function value, which would serve as a decision alternative indicator and facilitate screening of decision alternatives. This problem can be overcome by transforming all criteria measurements onto a common scale. The transformation of evaluation criteria to a common scale is also called *criterion standardization*. We concentrate here on transformation techniques that are applicable to deterministic criterion values (i.e., those that can be specified without considering uncertainty). The reader who is interested in criterion standardization techniques in which uncertainty must be taken into consideration should refer to probabilistic and fuzzy criterion values presented by Malczewski (1999).

Two common approaches to standardizations are linear and nonlinear. Formulas are called *linear* because they produce proportional (i.e., linear) transformations of raw input data. The simplest formula is called the *maximum score* procedure (also *ratio standardization*). The formula divides each raw criterion value by the maximum criterion value as

$$x'_{ij} = \frac{x_{ij}}{x_j^{max}},$$

(7.1)

where x'_{ij} is the standardized score for the ith decision alternative and the jth criterion, x_{ij} is the raw data value, and x_j^{max} is the maximum score for the jth criterion. The values of standardized scores can range from 0 to 1 and are linearly related to the raw data values. The formula in Equation 7.1 should be used *only* for the *benefit criteria* (the higher the raw data value the better the performance). An example of such a criterion might be "roads with good visibility." In the case of *cost criteria* (the lower the raw data value, the better the performance) the following formula for linear scale transformation should be used:

$$x'_{ij} = 1 - \frac{x_{ij}}{x_j^{max}}.$$

(7.2)

An alternative formula to the one presented in Equation 7.2 for cost criteria takes the inverse of the criterion outcomes. This results in rewriting the equation as follows:

$$x'_{ij} = \frac{x_j^{min}}{x_{ij}} , \qquad (7.3)$$

where x_j^{min} is the minimum score for the jth criterion.

As already indicated, the advantage of linear scale transformation is the proportionality between raw data values and standardized scores. This ensures that the relative distances between raw data values and standardized scores are preserved. A disadvantage of this procedure is that the lowest standardized score is not always equal to zero, which may cause some difficulties with the interpretation. Also, linear scale standardization produces difficult to interpret scores if the raw data values cover the range of negative and positive numbers. In this case, one should use nonlinear standardization.

In the nonlinear standardization procedure, the standardized value is computed for the benefit criterion by dividing the difference between a given criterion's raw data value and the minimum value of the value range:

$$x'_{ij} = \frac{x_{ij} - x_j^{min}}{x_j^{max} - x_j^{min}} . \qquad (7.4)$$

For the cost criterion, the standardized score is computed by dividing the difference between the maximum raw data value and a given raw data value from the value range:

$$x'_{ij} = \frac{x_j^{max} - x_{ij}}{x_j^{max} - x_j^{min}} . \qquad (7.5)$$

The advantage of this procedure is that the values of standardized criteria range precisely from 0 to 1. For each criterion, the worst standardized score always equals 0, and the best always equals 1. This is also true for negative raw criterion values. The disadvantage of nonlinear standardization is that unlike the linear scale standardization, this procedure does not produce standardized scores that are linearly related to raw scores. Hence, to preserve the proportions of raw scores in standardized scores, one should use linear scale standardization.

7.2 Transformation of Decision-Maker Preferences into Weights

An important step in the MCE approach is the articulation of decision-maker (DM) preferences in regard to criteria. MCE evaluation of decision alternatives frequently involves criteria of varying importance to the DM. The uneven importance of criteria may result from policies, established hierarchies, cause–effect relationships, and often subjective preferences. The preferences are expressions of one's values and in the MCE context represent the varying degrees of importance assigned to criteria. A common means of representing the DM's preferences are weights. *Weight* is a numeric amount assigned to an evaluation criterion, indicating its importance relative to other criteria in the decision situation. The weights are usually normalized, so that their sum for all *n*-criteria considered in a given decision situation equals 1. The larger the weight, the more important a given criterion is. It is important to understand that weights are influenced by differences in the range of variation in each criterion. The fact that the range of criterion value variation can influence the importance assigned to a criterion can be demonstrated in the following example: Consider an aquatic habitat restoration problem in which the restoration cost ranges from $1 million to $8 million per restoration site, and the effectiveness of restoration ranges from 85 to 100% restoration of the original habitat. Which of the two criteria would you deem to be more important? The general rule is that we are concerned with the perceived advantage of changing from the minimum level to the maximum level of each criterion outcome, relative to the advantages of changing from the worst to the best levels for other criteria under consideration. Consequently, one would deem the restoration cost as a more important criterion than restoration effectiveness, because the range of possible cost improvement from $8 million to $1 million is much greater (eight times, or 800%) than the range restoration effectiveness improvement from 85 to 100%.

Three common procedures for transforming DM preferences into numerical weights are *ranking*, *rating*, and *pairwise comparison*.

7.2.1 Ranking

Ranking is the simplest of all weighting techniques. It starts with the DM arranging criteria in an order of importance reflecting the DM's preferences. There are two common ways of doing this: *straight ranking* (1, the most important; 2, the second most important, etc.) or *inverse ranking* (1, the least important; 2, the next least important, etc.). Once the ranking is in place several procedures can be used to transform rank order information into numerical weights. Two of these procedures are presented below.

Rank sum procedure computes weights by using the following formula:

$$w_j = \frac{n - r_j + 1}{\sum_{k=1}^{n}(n - r_k + 1)}, \tag{7.6}$$

where w_j is the normalized weight (ranging in value from 0 to 1) for the criterion j, n is the number of criteria under consideration, and r_j is the rank position of the criterion.

Rank reciprocal procedure computes weights from the normalized reciprocals of a criterion's rank, using the following formula:

$$w_j = \frac{\frac{1}{r_j}}{\sum_{k=1}^{n}\frac{1}{r_k}}. \tag{7.7}$$

The meaning of symbols is here the same as in the rank sum formula.

ADVANTAGES AND DISADVANTAGES OF RANKING

The ranking approach to deriving criterion weights is attractive due to its simplicity. In practice, however, the number of criteria limits the practicality of ranking. The larger the number of criteria, the more difficult it is to arrive at a reliable ranking. One limit for the number of criteria, derived from psychometrics, is 7 ± 2. This means that, on average, humans are capable of discriminating, at the maximum, among nine different information elements (i.e., importance levels that can be assigned ranks). This limitation can be overcome by using hierarchical ranking schemes. Another limitation of ranking is its lack of theoretical foundations.

7.2.2 Rating

Rating requires the DM to estimate criterion weights on the basis of a predetermined scale. Commonly, a scale of 0 to 100 is used in concert with a *point allocation* approach. In this approach, the DM is asked to allocate 100 points across the evaluation criteria, pertinent to a given decision situation. The rational for allocation is simple: The more points a criterion receives, the greater its importance relative to other criteria. Assigning 0 points to a criterion is equivalent to ignoring it; conversely, assigning 100 points to one criterion is equivalent to ignoring all other criteria but this one. Individual criterion ratings can be normalized to fit the 0–1 scale and the requirement that the sum of all weights equals 1, using the following simple formula:

$$w_j = \frac{r_j}{100},\qquad(7.8)$$

where w_j is the normalized weight (ranging in value from 0 to 1) for the criterion j and r_j is the rating (a number from the range 0–100) assigned to the jth criterion.

ADVANTAGES AND DISADVANTAGES OF RATING

Rating is easy to explain using the analogy of distributing a fixed amount of money, on a priority basis, across a set of objectives represented by evaluation criteria. However, rating, similarly to ranking, lacks theoretical foundations, and its practicality is limited to a small number of criteria. The approach may be misused if the DM does not pay attention to the definition of criteria and ranges of criterion values.

7.2.3 Pairwise Comparison

The pairwise comparison technique, developed by Thomas Saaty in the 1970s and 1980s in the context of multiple criteria evaluation methods called analytical hierarchy process (AHP), represents a theoretically founded approach to computing weights representing the relative importance of criteria. Weights are not assigned directly but represent a "best fit" set of weights derived from the *eigenvector* of the square reciprocal matrix used to compare all possible pairs of criteria. The advantage of this technique is that information can be used from handbooks, regression output, or DMs/experts can be asked to rank order individual factors.

The technique comprises taking pairs of criteria (C_i and C_j) and asking two questions: (1) Which criterion is more important, C_i or C_j?, and (2) how much is the more important criterion relative to the lesser important criterion typically answered as "about the same" or "strongly more important," and subsequently scored on a 1–9 scale? Answers to these two questions are used to generate values in a square matrix A, where i is a row and j is a column. Because each factor is of equal importance to itself, the diagonal in A matrix is filled with 1's. If C_i (row element) and C_j (column element) are of equal importance, then a_{ij} (the value in the matrix A at the intersection of row i and column j) equals 1; if C_j is more important than C_i, then a_{ij} is set equal to the importance score and will be > 1; finally, if C_i is more important than C_j, then a_{ij} is set equal to the reciprocal of the importance score (i.e., 1/score) and will be < 1. The structure of the matrix A can be presented as follows:

$$
A =
\begin{pmatrix}
1 & a_{12} & a_{13} & \dots & \dots & \dots & \dots & a_{1n} \\
\frac{1}{a_{12}} & 1 & \dots & \dots & \dots & \dots & \dots & a_{2n} \\
\dots & & 1 & & & & & \\
\dots & & & 1 & & & & \\
\dots & & & & 1 & & & \\
\dots & & & & & 1 & & \\
\dots & & & & & & 1 & \\
\frac{1}{a_{1n}} & \frac{1}{a_{2n}} & \dots & \dots & \dots & \dots & \dots & 1,
\end{pmatrix}
\tag{7.9}
$$

where A is the reciprocal and square pairwise comparison matrix.

The entries a_{ij} in the matrix A are based on the 1–9 interval scale, with the following scale value meanings:

1—same importance
2—slightly more important
3—weakly more important
4—weakly to moderately more important
5—moderately more important
6—moderately to strongly more important
7—strongly more important
8—greatly more important
9—absolutely more important

More formally, the reciprocal and square pairwise comparison matrix can be defined in the following way: Let's assume the set of criteria

$$C_1, C_2, \ldots, C_n$$
$(C_i, C_j) \rightarrow$ a pair of criteria.

Then, the relative importance of each criterion can be represented by $n \times n$ matrix A, where $A = (a_{ij})$ $(i, j = 1, 2, \ldots, n)$.

Entries a_{ij} are defined by two rules:

Rule 1: If $a_{ij} = w$, then $a_{ji} = 1/w$.
Rule 2: If C_i is judged to be equally important as C_j, then $a_{ij} = 1 \wedge a_{ji} = 1$. In particular, $a_{ij} = 1$ for $i = j$.

The pairwise comparison procedure can be illustrated using the following example: Let us assume that three criteria are used in the decision problem of selecting a transportation improvement project:

C_1: access to jobs
C_2: air pollution emissions
C_3: annual passenger miles

C_3 is the most important criterion; it is weakly to moderately more important than C_2 (importance value, 4) and slightly more important than C_1 (importance value, 2). C_2 is the second most important criterion. It is slightly more important than C_1, which is the least important criterion. The following reciprocal and square pairwise comparison matrix can then be formed.

- Step 1: Specify square pairwise comparison matrix A.

	C_1	C_2	C_3
C_1	1	½	¼
C_2	2	1	½
C_3	4	2	1

$A =$

Notice that to specify fully reciprocal and square pairwise comparison matrix A, one needs to evaluate only $[n(n-1)]/2$ pairs of criteria. In the transportation project example, it is sufficient to evaluate $[n(n-1)]/2 = (3 \times 2)/2 = 3$ pairs of criteria to specify the matrix fully. The following pairs of criteria were compared: C_2 versus C_1, C_3 versus C_1, and C_3 versus C_2. The comparison outcomes were as follows: $a_{21} = 2$, $a_{31} = 4$, and $a_{32} = 2$. The remaining entries of matrix A were entered using the rules 1 and 2 (given earlier).

- Step 2: Synthesize judgments by summing the columns of matrix A.

	C_1	C_2	C_3
C_1	1	½	¼
C_2	2	1	½
C_3	4	2	1
	7	3.5	1.75

- Step 3: Normalize matrix A by dividing each column entry by the column's sum. To simplify the ratios, columns 2 and 3 were multiplied by 2 and 4, respectively.

	C_1	C_2	C_3
C_1	¹/₇	¹/₇	¹/₇
C_2	²/₇	²/₇	²/₇
C_3	⁴/₇	⁴/₇	⁴/₇

- Step 4: Compute the arithmetic average of each row in the normalized matrix A.

$$\Sigma\,(\text{row 1}) = \frac{1/7 + 1/7 + 1/7}{3} = 0.14$$

$$\Sigma\,(\text{row 2}) = \frac{2/7 + 2/7 + 2/7}{3} = 0.29$$

$$\Sigma\,(\text{row 3}) = \frac{4/7 + 4/7 + 4/7}{3} = 0.57$$

$$\Sigma\,(\text{row 1}) = 1 \text{ (note that the row averages sum up to 1.)}$$

The row averages provide an approximation of the *eigenvector* of the square reciprocal matrix A. The *eigenvector* of matrix A is an estimate of the relative weights of the criteria being compared. As we can see, based on three pairwise comparisons, the criterion access to jobs has the weight of 0.14, the criterion air pollution emissions has the weight of 0.29, and the criterion annual passenger miles has the weight of 0.57. Because the *eigenvector* of matrix A has the ratio scale property, we can interpret the weights by reasoning that the criterion annual passenger miles is twice as important as the criterion air pollution emissions, which in turn is almost twice as important as the criterion access to jobs.

Because individual judgments never agree perfectly, the degree of consistency achieved in the ratings is measured by a consistency ratio (*CR*) indicating the probability that the matrix ratings were randomly generated. The rule of thumb is that a *CR* less than or equal to 0.10 indicates an acceptable reciprocal matrix A, and ratios over 0.10 indicate that the matrix should be revised. Revising the matrix comes down to (1) finding inconsistent judgments regarding the importance of criteria, and (2) revising these judgments by comparing again the pairs of criteria judged inconsistently.

- Step 5: Compute CR. This computation is carried out in a few substeps.
- Step 5a: Determine the weighted sum vector by multiplying matrix A by the vector of criterion weights (each column is multiplied by the corresponding criterion weights, and the products are summed over the rows).

$$\begin{bmatrix} 1 & \frac{1}{2} & \frac{1}{4} \\ 2 & 1 & \frac{1}{2} \\ 4 & 2 & 1 \end{bmatrix} \times [0.14 \quad 0.29 \quad 0.57] = \begin{bmatrix} 0.428 \\ 0.86 \\ 1.71 \end{bmatrix}$$

- Step 5b: Determine the consistency vector by dividing the weighted sum vector by the criterion weights.

$$\begin{bmatrix} 0.428 \\ 0.86 \\ 1.71 \end{bmatrix} \div \begin{bmatrix} 0.14 \\ 0.29 \\ 0.57 \end{bmatrix} = \begin{bmatrix} 3.06 \\ 2.97 \\ 3 \end{bmatrix}$$

- Step 5c: Compute the λ term (the average value of the consistency vector).

$$\begin{bmatrix} 3.06 \\ 2.97 \\ 3 \end{bmatrix} \div 3 = \frac{3.06 + 2.97 + 3}{3} = 3.01$$

- Step 5d: Compute the consistency index (CI). The calculation of CI is based on the observation that λ is always greater or equal to the number of criteria (n), and $\lambda = n$ if the pairwise comparison matrix A is a consistent matrix. Accordingly, $\lambda - n$ can be considered as a measure of the degree of inconsistency. This measure can be normalized as follows:

$$CI = \frac{\lambda - n}{n - 1} = \frac{3.01 - 3}{3 - 1} = 0.005$$

CI provides a measure of departure from consistency.

- Step 5e: Compute the CR defined as

$$CR = \frac{CI}{RI} = \frac{0.005}{0.58} = 0.009,$$

where RI is the random index representing the consistency of a randomly generated pairwise comparison matrix. Tabulated values of RI can be found in the AHP literature (Saaty, 1980). For example, the RI values for the number of evaluation criteria ranging from 1 to 8 are as follows:

n	1	2	3	4	5	6	7	8
RI	0	0	0.52	0.89	1.11	1.25	1.35	1.4

The value of RI depends on the number of criteria being compared. The value of CR (0.009) falls much below the threshold value (0.1), and it indicates a high level of consistency. Hence, we can accept the weights.

ADVANTAGES AND DISADVANTAGES OF PAIRWISE COMPARISON

Pairwise comparison, in contrast to ranking and rating, has a solid theoretical foundation based on ratio scale judgments about pairs of criteria and the properties of a reciprocal matrix of pairwise comparisons. The advantage of this technique is that information can be used from handbooks, regression output, or DMs/experts can be asked to rank order individual factors. The disadvantage of pairwise comparison is the large number of judgments that must be made when the number of criteria is large.

7.3 Decision Rules in MCE

A *decision rule* is a procedure for ordering alternatives from most to least desirable. The use of a decision rule may facilitate (1) selection of the most desirable alternative, (2) sorting of alternatives into classes arranged into a priority order, and (3) ranking of alternatives from best to worst. Decision rules provide the basis for selection, sorting, and ranking by integrating the data on alternatives and DM's preferences into an overall assessment of the alternative. This assessment is often expressed by one score, also called the *overall appraisal score*. In the MCE approach, the overall appraisal score is the value of a function that aggregates the outcomes of a decision alternative over all evaluation criteria with the DM's preferences. This is why decision rules in the MCE context are also called *combination functions*. Many decision rules have been proposed in the MCE literature (Voogd 1983). We present an overview of two popular rules: *weighted linear combination* and *ideal point*. A more comprehensive overview of MCE decision rules can be found in Malczewski (1999). Each rule is based on a different rationale, leading to the computation of an overall appraisal score.

7.3.1 Weighted Linear Combination Decision Rule

The weighted linear combination (WLC) decision rule has been widely used for its simplicity. A final appraisal score e_i for each alternative i is computed by multiplying the jth criterion importance weight w_j by the standardized outcome score of alternative i on criterion j. The assumption for using the WLC decision rule is that evaluation criteria are preferentially independent (the importance attached to one criterion is independent from the importance attached to other criteria).

$$e_i = \sum_{j=1}^{n} w_j \bullet r_{ij}, \quad i = 1, \ldots, m \tag{7.10}$$

$$i \rightarrow \text{decision options} = m$$
$$j \rightarrow \text{criteria} = n$$

In the matrix notation, the preceding formula can be presented as follows:

$$\begin{bmatrix} e_1 \\ : \\ : \\ e_m \end{bmatrix} = \begin{bmatrix} r_{11}\ldots\ldots\ldots r_{1n} \\ : \\ : \\ r_{m1}\ldots\ldots\ldots r_{mn} \end{bmatrix} \times [w_1 \ldots w_n],$$

where e_i is the evaluation score (also called appraisal score) computed for decision alternative i and w_j is the weight representing the relative importance of criterion j.

Weighted linear combination has been implemented in GIS packages such as Arc-GIS, IDRISI, SPANS, and CommonGIS. The decision rule can be operationalized in any GIS software with overlay capabilities. The WLC was implemented in ArcGIS as the Weighted Overlay tool, and it can be found in the Overlay toolset belonging to the Spatial Analyst toolbox.

7.3.2 Ideal Point Decision Rule

The *ideal point decision rule* calculates the final appraisal score for each decision alternative based on the separation of combined alternative outcomes from the ideal point. The ideal point represents a hypothetical alternative that comprises the most desirable outcomes for the evaluation criteria. The nadir represents a hypothetical alternative that comprises the least desirable outcomes for evaluation criteria. The alternative that is closest to the ideal point, and at the same time farthest from its nadir, is the best alternative under this decision rule.

The decision rule can be computed with the following procedure:

1. Calculate standardized criterion scores using either the linear standardization formula (below) or the score range standardization formula.

$$X'_{ij} = \frac{X_{ij}}{X_j^{\max}}$$

For negative values, use the score range standardization procedure.

2. Calculate weighted standardized criterion scores.

$$V_{ij} = W_j \bullet X'_{ij}$$

3. Identify positive-ideal and negative-nadir solutions.

$A^* \to$ ideal

$A^* = \{v^*_1, v^*_2, \ldots, v^*_j, \ldots, v^*_n\}$

$A^* = \{(\max v_{ij} | j \in J_1),(\min v_{ij} | j \in J_2) | i = 1, \ldots, m\}$

    J_1: set of benefit criteria

    J_2: set of cost criteria

$A^- =$ nadir

$A^- = \{v^-_1, v^-_2, \ldots, v^-_j, \ldots, v^-_n\}$

$A^- = \{(\min v_{ij} | j \in J_1),(\max v_{ij} | j \in J_2) | i = 1, \ldots, m\}$

4. Calculate separation measures from ideal – S^* and nadir – S^-.

$$S^* = \sqrt{\sum_{j=1}^{n}(v_{ij} - v^*_j)^2}, i = 1, \ldots, m$$

$$S^- = \sqrt{\sum_{j=1}^{n}(v_{ij} - v^-_j)^2}, i = 1, \ldots, m$$

5. Calculate the index of similarities to ideal.

$$C^*_i = \frac{S^-_i}{S^*_i + S^-_i}, i = 1, \ldots, m$$

$$0 \le C^*_i \le 1$$

$C^*_i = 0$, when $A_i = A^-$

$C^*_i = 1$, when $A_i = A^*$

6. Create preference order of decision options, choose an option with the maximum C^*_i or rank the options according to C^*_i in the descending order.

Ideal point method relies on the notion of the best possible set of criterion scores as influenced by three aspects: (1) the ideal, (2) its nadir (i.e., worst combination of criterion scores), and (3) the distances from each option to the ideal and the nadir.

Ideal point decision rule provides complete, interval scale ranking of decision alternatives. This means that the relative distance of each alternative to the ideal point can be computed. This decision rule avoids the restrictive assumption of independence among the evaluation criteria—made by additive and value/utility function-based decision rules. This makes the ideal point decision rule an attractive approach to decision problems in which the independence among criteria is difficult to test. This is especially true in spatial decision problems involving geographically influenced interdependencies among the evaluation criteria.

EXAMPLE: IDEAL AND NADIR

The concepts of ideal and nadir can be explained by the following simple numerical example. Let's have rows represent geographically distributed options in a decision analysis (as e.g., potential corridors for development), and columns represent the evaluation criteria. We will assume that "high values are better" for each of the criteria, which means that all criteria are the benefit criteria.

$$
\begin{array}{cccc}
12 & 8 & 11 & 5 \\
10 & 7 & 10 & 4 \\
16 & 5 & 12 & 2
\end{array}
$$

For this simple decision table of three decision options and four criteria the ideal is [16,8,12,5] and the nadir is [10,5,10,2]; that is, 16, 8, 12 and 5 are the highest values over the four criteria (in each of the four columns), whereas 10, 5, 10, and 2 are the lowest values over the four criteria. Note, then, that the ideal represents the best of all options for all criteria, and the nadir, the worst of all options for all criteria.

Consequently, in the ideal point method, each of three options is compared to the ideal and the nadir. The data value distance measured to the ideal and the nadir, calculated for each analysis option, are then aggregated into relative closeness measures. The *closeness measure* expresses how close each option is to the ideal and, conversely, how far it is from the nadir. The options are then ordered/ranked, beginning with the one closest to the ideal and farthest from the nadir, and ending with the one farthest from the ideal and closest to the nadir. The maximum possible evaluation score (the relative closeness measure) is 100, and the minimum possible is 0.

The ideal point decision rule was implemented in GeoChoicePerspectives (Jankowski and Nyerges 2001b), a spatial decision support system software that comprised a tightly coupled GIS component (ArcView 3.x), an MCE component called ChoiceExplorer, and a group decision process component called ChoicePerspectives. This rule was also implemented in CommonGIS (Andrienko, Andrienko, and Jankowski 2003).

The MCE transformations are relevant to situations involving (1) sorting or (2) rank ordering of planning scenarios. Sorting of plans into acceptable and nonacceptable may be accomplished on the basis of a final appraisal score by applying, for example, a cutoff value. The MCE transformations are also relevant to improvement programs when limited funds preclude funding all worthy improvement projects. In such situations, a ranking of projects, produced with help a decision rule, can become the basis for a funding selection decision.

A potential criticism of MCE transformations in the context of decision support in planning and improvement programming concerns the explicit representation of preferences by numerical weights. The weights introduce an element of bias and subjectivity. Because the improvement programming decisions are frequently the subject of political scrutiny, in which a systematic bias introduced at the technique level may render the results of evaluation an easy target of critique for those who disagree with the final selection of improvement projects, one may choose to omit the weights altogether and still be able to use the MCE transformations. Mathematically, this amounts to dividing the value of 1 by n-weights and assigning each weight the value of $1/n$. The meaning of such an assignment is to make each evaluation criterion equally preferable. There are also other, multiple-criteria decision-making methods, in which preferences are captured implicitly by either asking the DM to evaluate trade-offs among the criteria (Jankowski, Lotov, and Gusev 1999) or inferring *if ... then ...* -type decision rules from decision examples (Pawlak and Slowinski 1994). The latter approach is attractive in situations where people prefer to make exemplary decisions instead of revealing their preference structure in terms of criterion weights or trade-offs, and cannot always explain them in terms of specific parameters. From this point of view, the idea of inferring preference models from exemplary decisions provided by the DM is very attractive. However, the exemplary decisions may be inconsistent because of some additional aspects that are not included in the considered family of criteria, and because of hesitation on the part of the DM. These inconsistencies cannot be considered as simple errors or as noise. They can convey important information that should be taken into account in the construction of the DM's preference model.

7.4 Sensitivity Analysis in MCE

It is important to realize that despite our best efforts to compute accurate criterion values, provide a complete set of relevant evaluation criteria, and elicit true preferences of the DMs, the data and assumptions of any MCE analysis are subject to change and error. *Sensitivity analysis* (SA) is the investigation of these potential changes and errors, and their impacts on the MCE solutions. A methodology for conducting an SA is a well-established requirement for any modeling study. One cannot truly claim to have a good understanding of the stability of model solution on plausible changes in the input parameters without conducting an SA. In regard to MCE, this affects particularly investigation of the effects of shifting priorities and changing evaluation criteria, and alters the set of decision options under consideration.

Priorities, often expressed by numerical weights in MCE techniques, may shift as a result of changing information, political situation, persuasion, or simply DM's views. The purpose of performing an SA is to check how stable the ranking of options is in reaction to changes in criterion weights. If small changes in criterion weights have no influence on the ranking of options, one may have more confidence in the stability of one's ranking. If, on the other hand, small changes in weights change the ranking, one may want to reexamine one's preferences or simply accept the fact that the small change in preferences may shift the order of options. One way to get a quick understanding of the influence of criterion weight on the ranking scores is to give it a full weight of 100 (and push others to 0), and see what happens to the overall appraisal score for each decision alternative.

Evaluation criteria may be dropped or new criteria may be added for similar reasons, as in case of criterion priorities. The simplest form of SA in regard to criteria is to observe the effects of deleting or adding criteria on the ranking of options. One can also use a more systematic approach by deleting one criterion at a time, and checking the effect of deletion on the ranking. By repeating this step for each criterion one can identify the "weak" criteria that have little effect on the overall ranking. Such criteria potentially can be eliminated without altering the solution. A more complex approach to deciding which criteria are weak and can be eliminated involves global SA, which decomposes the variance of the output of the MCE process into a variety of explanatory factors. An example of a global SA method is the extended Fourier amplitude sensitivity test (FAST). The extended FAST method uses *first* and *total* order indices as SA measures (Crosetto and Tarantola 2001). The *first-order sensitivity index* is defined as a fractional contribution to the variance of MCE model output (e.g., option ranking) due to the uncertainty of a given input parameter treated independently from other parameters. The *total order index* represents the overall contribution of a given parameter (e.g., criterion weight), including its interactions with other parameters. Computation of the indices requires a large number of rank order calculations performed with weight vectors derived from the DM's weight distribution functions (Saisana, Saltelli, and Tarantola 2005).

Similar to changing the set of evaluation criteria in an effort to identify noncritical criteria, one can change the set of decision options and observe changes in option ranking, while adding or deleting an option. Often, two decision options score close to each other. One can then investigate changes needed in a given criterion weight or in criteria scores to bring two options to a tie. If the changes are small, then the information is useful for making a choice between the top-scoring and the runner-up options.

So far we have discussed GIS and MCE analysis methods, which can help to address user information needs. In the following section we present an example of how these methods may be applied in practice in the context of finding suitable sites for a wastewater treatment facility, in which each suitable site is a project option.

7.5 MCE of Site Alternatives for Green County Wastewater Facility

Back to Chapter 3, we indicated that the Green County GIS project was about area planning, in which several sites were identified for possible wastewater facilities. In the following section we present the site selection analysis for a Green County wastewater facility, but with a focus on MCE, as described earlier. We start the analysis by discussing multiple perspectives of stakeholders interested in building a wastewater facility for the Green County community. In Chapter 3, the analysis was motivated by policy guidelines from a county handbook. However, it is important to incorporate into a GIS project the stakeholders' perspectives and how their concerns can be interpreted as site evaluation criteria. Eliciting stakeholder concerns as part of public participation processes improves the chances that data analyses are less likely to be challenged in court; people's voices will have been heard.

In this section, we assume that a preliminary analysis has been performed; thus, available parcels have been identified, as in Chapter 3. In a conventional data analysis, as in Chapters 3 and 6, an important point to remember is that if one or more of the DMs wanted to change the assumptions about what constitutes a "suitable" parcel, or even just explore hypothetically a different criterion consideration, then we could not easily return to the full dataset to incorporate that additional perspective. We have shown in Chapter 6 how information needs can be addressed by extending the simple GIS workflow approach to incorporate process, scenario, change, and impact models. The interactive MCE extends the preliminary analysis by combining information about suitable parcels with stakeholder preferences, using the decision rules presented in section 7.3, and supports a test of the robustness of site ranking with help from an SA.

7.5.1 Stakeholder Perspectives on Siting a Green County Wastewater Facility

The decision problem of which site to choose for the wastewater facility is considered by the Green County Facility Siting Panel representing different stakeholder groups. The panel balances strong values for environmental protection and stewardship, and economic development—a foundational perspective for integrated resource management. Generally, the panel prefers a long-term, holistic approach rather than a short-term approach to improving and managing water resources, and also a commitment to implementing long-term maintenance and sustainability of water resource projects. After all, a wastewater project is indeed a water resource project, influencing the quality of water in a community.

How panel members work together, how the panel spends available money and how it communicates with the public are other strong values held by the panel. Some panel members feel strongly that the panel should function as a partnership, working collaboratively beyond individual interests to address the broader picture of environmental stewardship for future generations. Cost-effectiveness is a strong value, expressed primarily in the sense of getting the maximum return for dollars spent, which includes the leveraging of funds from other sources to produce greater results. Some panel members see an important need to involve the public early in the process, so that people know what the panel is doing and can be encouraged to become stewards, and so the panel can benefit most from the public's points of view. The panel has agreed to listen to members of user groups who voice different values for decision making. An open decision process is being encouraged at all times. In addition to the panel's perspective specific stakeholder group perspectives on the decision problem reveal concerns that ought to be addressed in the site selection process.

7.5.1.1 A Regulatory/Resource Agency Perspective

Representatives of the regulatory/resource agencies express a range of values. Generally, the agencies show a strong concern for protecting the environment and preventing pollution in the river according to their respective agency mandates and missions. A number of the agencies believe that controlling pollution should be the first priority for the panel. Some believe that the best approach is to balance potentially conflicting needs, striving for the best return on the investment, and to use the panel's work as a catalyst for more work later as values are shared by a number of the agencies.

Members of one agency suggested that environmental sustainability as the fundamental reason for siting a facility may lead to the possible siting of other, similar facilities. Another value that some agency members feel the panel should rank high in its criteria for selecting projects is attention to the health risk to people. One agency ranks the health risk priority ahead of the health of the ecosystem and fish, which it in turn ranks ahead of economic feasibility. According to another agency, overall water quality is the primary value that should guide the panel, followed by benefits to the public. Impacts on fisheries and on restoring the bays and rivers in close proximity to a level that will support fish and other aquatic life are values expressed by still other agencies. The effect of the panel's work on navigation and commerce is of concern to other agencies.

The value that focuses on controlling pollution encourages a regulatory/resource person to recommend siting the facility as far from the river as possible. Selecting projects that establish optimum conditions for conservation of water resources and keep the flood zone free from potential overflow hazards is what this group believes should be pursued.

Two evaluation criteria derive directly from this perspective: (1) maximizing the distance of the facility from the river, and (2) selecting sites outside the floodplain zone.

7.5.1.2 An Elected Official's Perspective

The elected officials show a strong appreciation for the environment and its importance to people and to the region's quality of life. Clean water is one of the important values expressed by the elected officials, with one official citing the importance of clean water for children. Correcting environmental problems, and the sources of those problems, is also an important value, as long as doing so does not create other problems. In fact, the work of the panel is seen as an opportunity to look for creative solutions, some of which may solve multiple problems.

Gaining the most long-term value for money spent is another strong value of the elected officials. The challenge may be in finding ways to get the most out of the dollars available, which includes joining with others to obtain more resources, if possible. The City Planning Department suggests that it would be less expensive to site a facility within the city limits. Least expensive are the vacant parcels identified by the County Assessor.

Individual elected officials show interest in developing public safety and odor control, keeping the facility as far from residential housing as possible. Maintaining a safe distance from housing ensures that the assessed value of nearby homes does not drop.

Two evaluation criteria, in addition to those already established according to the regulatory agency perspective, can be derived from the elected officials' perspective: (1) sites should be contained in vacant parcels and be within the city limits, and (2) the distance between candidate sites and residential properties should be maximized.

7.5.1.3 An Engineering Consulting/Academic Perspective

The consulting/academic individuals focus to a great degree on planning ahead, picking priorities, and choosing projects that help us understand the external problems within the community, so that they can avoid such problems in the future, if at all possible. One individual voiced a strong need for a regional plan for siting, noting that a project-by-project approach will not do the job of enhancing water resources. Another individual advised the panel to fit its work into a total scheme and be sure to know the outcome it wants to accomplish. Because many things can be done, the panel must set priorities. As a way to set priorities, the consulting/academic individuals value risk assessment, with a focus on clear risks to people; maintenance of existing resources, while looking for other opportunities; and the ability of projects to sustain themselves beyond the panel's work. Several of the individuals saw a benefit in developing projects that are transferable to other locations, either in this region or across the nation.

Weighing costs against benefits and choosing the most effective projects for the money spent are important values to the consulting/academic individuals. Spending a lot of money to clean soils that lend minor benefits to cleaning wastewater, for example, would only indirectly influence these values. One individual advocated putting benefits before costs in seeking real improvement and protection of resources.

Another value cited by this group is to be sensible; to build adaptive management and feedback loops into projects; to make efforts to avoid historical areas; and to set up conditions that foster clean water everywhere. This group favors putting the facility as close to largest wastewater junctions as possible, thereby constraining wastewater flows over long distances. It also favors putting the facility on a parcel whose elevation is lower than 365 feet to reduce pumping costs. Larger parcels are favored over the smaller ones.

Four new evaluation criteria come out of this perspective: (1) sites should not overlap with historical areas and (2) they should be as close as possible to largest wastewater junctions, (3) have parcel elevation lower than 365 feet, and (4) maximize the parcel size.

7.5.1.4 An Environmental Group Perspective

Representatives of environmental groups feel strongly about preventing pollution and reducing risks to people and the environment, restoring areas to a greatly improved state, and finding ways to involve the public. An important value is pollution prevention and the protection of living things, including people. Generally, priorities should be based on reducing the greatest risks to human and environmental health. Protection of salmon is important in part because of its economic contribution to the region's quality of life. Using cost–benefit analysis to help set priorities, focusing on the control of toxic waste, fully understanding problems before applying new technology, and developing long-term approaches to long-term problems are other, related values expressed by individuals.

One individual called for recycling where the biggest drawdown on water is occurring, to replace nonrecycled with recycled water when and where possible. Another individual advised the panel to develop a bold vision, suggesting that the least natural parcel in the areas (e.g., a brownfield) should be examined as a top priority.

Involvement of the public in some way is also important to the environmental representatives. Public access and enjoyment, as long as it does not impact getting the job done, leads to education and better understanding of the problems facing us. The panel should promote stewardship in a sense that takes into account future uses of the area. A public process also helps the panel to establish values in making decisions and to ease implementation.

The values and concerns voiced by the representatives of environmental group point to some of the same criteria that we have already identified, so, for now, no new criteria are added to the list.

7.5.1.5 A Business/Community Leader Perspective

Business and community leaders express a range of values that the panel should use in guiding its decisions. To varying degrees, many of these leaders acknowledge that the panel should try to site a facility, with the least overall impact to the community. The panel should balance economic values with social values. Reasonableness and an eye to multiuse of the facility may be important in finding a balance between potentially conflicting needs. Yet efforts to restore the water resources, remediate land use problem areas, and eliminate sources of pollution are very important to these leaders. In fact, eliminating sources of pollution is a higher priority than employment opportunities for many of the leaders.

A major value of the business and community leaders is to ensure protection for public health and the environment. Some believe that clean water should be the panel's goal by getting rid of the sources of pollution, including discharges from ships and boats, or by making it more costly for polluters to pollute. Public health is also important, particularly because pollution affects the quality of fish and shellfish. Several leaders used the phrase "fishable/swimmable," to summarize their definition of clean water. Thus, keeping the facility away from the river is a major consideration for this group.

The business and community leaders are interested in effective, optimum use of the panel's money. Their focus is on getting the best value for the limited funds available, and using the money to get people to work together. Spending excessive amounts of money to get infinitesimal results is not valued, but doing the job right the first time is valued. The group wants the sites to be located as close as possible to the existing roads, in order to use the existing infrastructure. This is yet another criterion.

7.5.2 Stakeholder Perspectives and Objectives for Interpreting Criteria Data Values

Various stakeholder perspectives contain personal and/or organizational values that sometimes get articulated as specific concerns and recommendations that can be adopted as evaluation criteria. As such, based on those stakeholder perspectives, we identified nine criteria. The criteria present specific user information needs in terms arriving at stakeholders' values for each identified site. To determine these criteria data values, we need to compute them. But how? Let us use the framework introduced in section 6.2. Notice that many of the criteria correspond to spatial relationship–data transformation combinations presented in Tables 6.2–6.5. Below we describe the nine criteria and the corresponding GIS data analysis functions used to compute the criteria data values.

1. *Identify parcels that are below 365 feet.* Table 6.4; containment/polygon → polygon; clip from the polygon layer the available parcels with areas below 365 feet. Also, can use overlap/polygon → polygon; overlay intersect polygon elevation layer with parcel layer.

2. *Locate parcels that are within the city.* Table 6.4; overlap/polygon → polygon and containment/polygon → polygon; first, overlay union of available parcels with the city boundary polygon layer; next, select from the result layer the polygons representing parcels within the city boundary.

3. *Find large-size properties.* No specific transformation is needed to meet this criterion; however, the parcels can be found easily through an attribute query operation.

4. *Identify parcels that are not in the historical district.* Table 6.4; (No) overlap/polygon → polygon and containment/polygon → polygon; first, overlay union of available parcels with the historical district polygon layer; next, select from the result layer the polygons representing parcels outside the historical district boundary.

5. *Minimize the distance from parcels to the closest wastewater pipe junction.* First we observe that there is a GIS data analysis operation corresponding to the proximity/point → polygon combination in Table 6.2. In this instance we need to abstract polygon to point, which can be the polygon's centroid. We then look at the combination proximity/point → point, which offers point distance operation. Using it, we can find distances between each parcel centroid and each point representing the wastewater pipe junction.

6. *Maximize the distance from parcels to the closest river.* Similar to the operation in criterion 5, we first need to transform parcel polygons to points (centroids), then the combination proximity/point → line affords as near operation, which can compute the shortest Euclidean distance between each point and the corresponding location on the line. We can then sort the distances in increasing order to find parcels that are farthest away from the river.

7. *Minimize the distance from parcels to the closest road.* Apply the same combination as for criterion 6.

8. *Maximize the distance from parcels to residential properties.* Apply the same combination as for criterion 5, except that here we need to transform from polygon to points (centroids) both layers: parcels and residential properties.

9. *Find parcels located outside the floodplain.* Apply the same combination as for criterion 4.

7.5.3 Criteria Data Values Used for Site Ranking

The GIS operations applied to input data layers result in computing values for the criteria. The values can be arranged into the decision table, with raw records representing candidate sites and columns representing criteria. Before one can apply any of the decision rules presented in section 7.3, criteria data values must be standardized (section 7.1), and preferences in regard to evaluation criteria must be enumerated (section 7.2). Let us assume that the criteria data values have been standardized with the nonlinear standardization formulas (Equations 7.2–7.3, section 7.1), and that we have used the rating formula (Equation 7.8) to arrive at criterion weights (section 7.2). The weights are given in Table 7.1.

The weights in Table 7.1 represent one specific perspective that seems to be close to the business/community leader perspective protecting residential properties (highest weight = 15) and promoting cost-related criteria (parcels in the city, distance to wastewater pipe junction, distance to roads).

TABLE 7.1. Weights for the Green County Criteria

Criterion	Weight
Identify parcels that are below 365 feet.	9
Locate parcels that are within the city.	11
Find large-size properties.	11
Identify parcels that are not in the historical district.	10
Minimize the distance from parcels to the closest wastewater pipe junction.	12
Maximize the distance from parcels to the closest river.	11
Minimize the distance from parcels to the closest road.	13
Maximize the distance from parcels to residential properties.	15
Find parcels located outside the floodplain.	8

ChoiceExplorer 1.2 - Untitled			
File Tools Vote Help			
Criteria	**Weight**	**Options**	**Score**
Elevation	09	64	68
Area_1	11	101	62
Not_his_di	11	98	62
Within_cit	10	246	61
Junction_d	12	231	61
River_dist	11	225	61
Road_dist	13	211	61
Resid_dist	15	106	61
Flood_dist	08	102	61
		100	61
Weighting: RAT (EQL) Aggregation: IP			

FIGURE 7.1. Criteria weights and the ranking of candidate sites.

We use the weights and present the ranking for the 10 top-ranked sites (out of the larger set of 283 candidate parcels in Green County, these are the option identifier numbers) on the right-hand side of Figure 7.1. The ranking was obtained with the ideal point decision rule (section 7.3). Site option 64 is the top-ranked location, with a final appraisal score of 68. The next-best site 101 has a final appraisal score of 62, which is almost 10% less than the top score. Looking at the ranking, we may be ready to conclude that site 64 should be recommended to the panel for the wastewater facility location. Before we can make the recommendation, we should check the sensitivity of the ranking. Let us assume that we are confident about criteria and criteria data values: We do not expect some of the criteria in the set to become irrelevant, and similarly we do expect our criterion values to change in the future. However, we are not so certain about the priorities, which can shift up or down. It is always difficult to predict how large such shifts can be, and a 20% shift up or down seems to be a plausible guess. This would mean that the highest-priority criterion "Resid_dist" (distance from parcels to residential properties) could change the weight by ± 3 from its current weight of 15.

We can test whether such changes in weights will result in a change of the ranking, especially for top-ranked site 64. The result of such change is presented in Figure 7.2.

The weight for the criterion "Resid_dist" was increased from 15 to 18, and the other weights were adjusted proportionally. This resulted in site 64 staying still at the top of

ChoiceExplorer 1.2 - Untitled			
File Tools Vote Help			
Criteria	**Weight**	**Options**	**Score**
Elevation	09	64	65
Area_1	10	184	60
Not_his_di	10	231	60
Within_cit	09	225	60
Junction_d	12	211	60
River_dist	11	210	60
Road_dist	13	199	60
Resid_dist	18	75	60
Flood_dist	08	46	60
		185	59
Weighting: USER (EQL) Aggregation: IP			

FIGURE 7.2. The result of change in weights on the site ranking.

the ranking, albeit with a smaller final appraisal score of 65. However, the order of the lower-ranked sites changed (cf. Figures 7.1 and 7.2). This indicates to us that top-ranked site 64 is stable, and that the ranks of other sites may change due to shifting weights. In light of this result of the SA, we recommend site 64 to the panel.

7.6 Summary

MCE techniques are an extension of basic GIS-based decision analysis. MCE can add to the development of an integrated decision environment by increasing the interactivity of decision analysis. We extended the basic data analysis approach with MCE that enables systematic evaluation of trade-offs among suitable choice options. We transformed raw criterion data values into standardized scores, so that we could compare data values across attributes. Preferences provide a way of assigning weights using ranking, rating, and pair-wise comparision. Ranking implements weights using a low-to-high sequence order within the attribute set. Rating assigns a score, because the unit value describing the weights is meaningful. Pair-wise comparison derives criterion weights based on the systematic examination of the importance of each criterion in a set of evaluation criteria under consideration. Decision rules for the weighted linear combination and the ideal point were presented. A linear weighted combination assumes that we can add and multiply weights by scores to compute an overall outcome to establish priority. Ideal point works within the data range of each attribute in the dataset, comparing each attribute to an "ideal" data value to construct an overall ideal option. Many of the data analysis techniques presented in Chapters 6 and 7 are applied to case studies in the chapters to come.

7.7 Review Questions

1. How does the analysis of user information needs motivate use of spatial multiple criteria evaluation techniques in decision analysis?

2. Why do we want to transform raw criterion data values into standardized criterion scores?

3. What is(are) the ramification(s) of using a linear scale transformation procedure as opposed to a nonlinear (e.g., score range) transformation procedure for standardizing scores?

4. What is the significance of DM preferences, and what is the major conceptual difference between ranking and rating?

5. What is a reciprocal pairwise comparison matrix, and how would one use it in MCE?

6. What are the weighted linear combination and ideal point decision rules?

7. What are advantages and disadvantages of using weighted linear combination and ideal point decision rules for establishing option priorities?

8. What is SA, and how do we use it to clarify option priorities?

9. Why are stakeholder perspectives important to consider? What are the similarities and differences among the stakeholder perspectives for siting the Green County wastewater facility?

10. How are personal/organizational values reflected in the stakeholder perspectives and used in MCE to rank order candidate site locations for the wastewater facility?

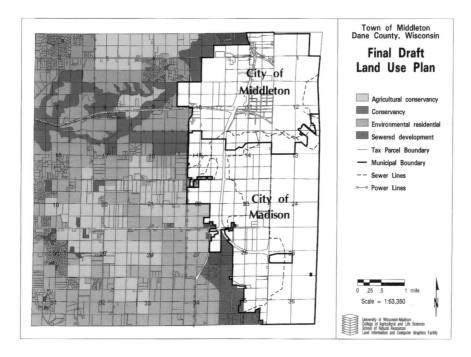

PLATE 2.1. Land use plan for Middleton Township, Wisconsin. From Niemann (1989).

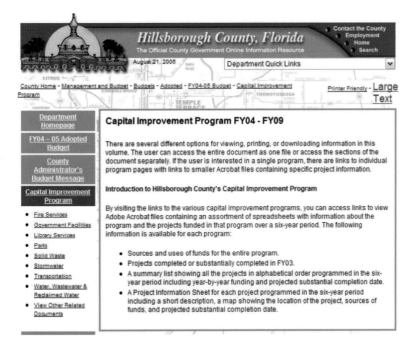

PLATE 2.2. Hillsborough County capital improvement program project tracking system. From Hillsborough County (2006). Copyright 2006 by Hillsborough County. Reprinted by permission.

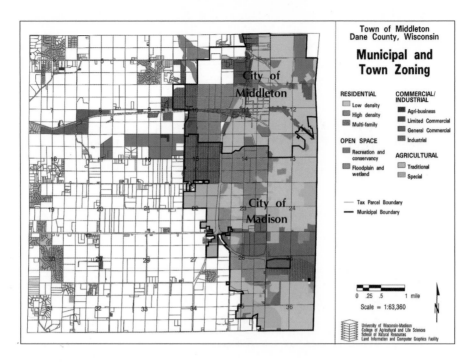

PLATE 2.3. Zoning map for Middleton Township, City of Middleton, and a portion of Madison, Wisconsin. From Niemann (1989).

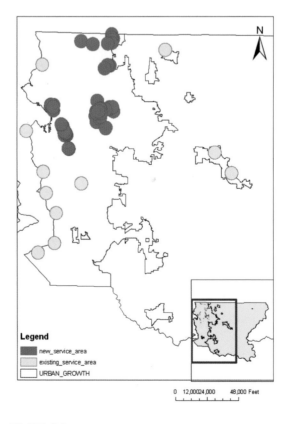

PLATE 3.1. Map depicting an expanded service area.

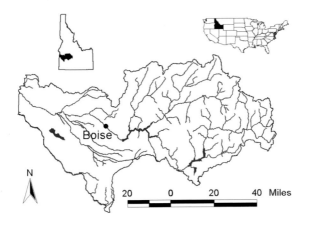

Boise River Basin

PLATE 9.1. Boise River Basin study area in southwestern Idaho.

PLATE 9.2. A facilitated water resource conjunctive administration meeting in Boise, Idaho, consisting of a single-user version of WaterGroup GIS.

PLATE 9.3. A facilitated water resource conjunctive administration meeting in Boise, Idaho, consisting of a multiple-user version of WaterGroup GIS.

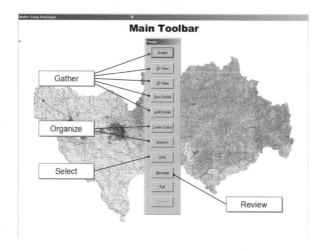

PLATE 9.4. Toolbar from WaterGroup GIS.

PLATE 9.5. Aerial imagery of 1-meter resolution.

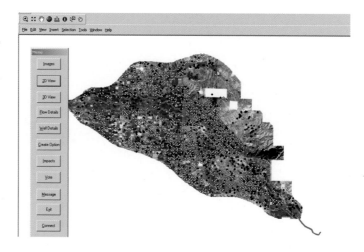

PLATE 9.6. A 2D map display of wells for a significant portion of the basin. Three-dimensional map displays depict the well depths (see Plate 9.7). The displays can be tilted and rotated to provide a better sense of the distribution of well depths across the basin.

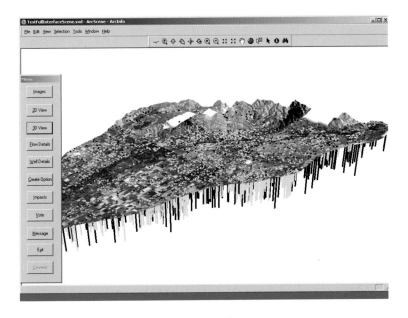

PLATE 9.7. A 3D visualization depicting well depth.

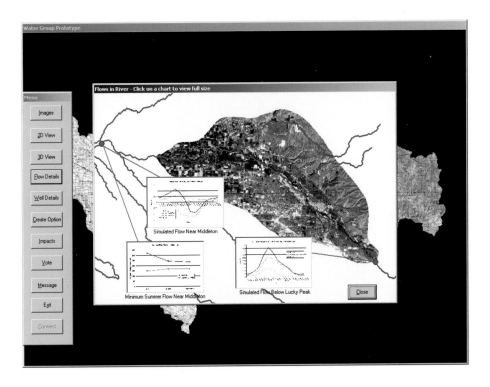

PLATE 9.8. A 2D map with flow details highlighted with a hot link.

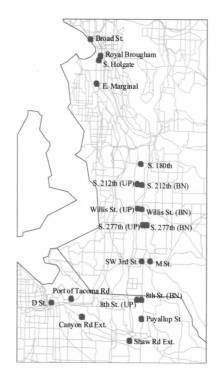

PLATE 10.1. FAST freight mobility (railway-grade separation) sites in a decision experiment.

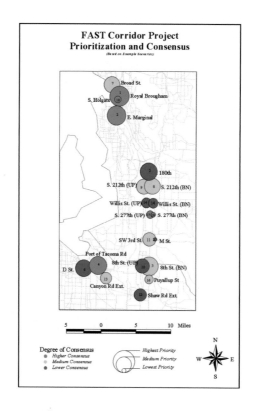

PLATE 10.2. Consensus map of FAST program. From Jankowski and Nyerges (2001a). Copyright 2001 by Taylor & Francis Ltd. Reprinted by permission.

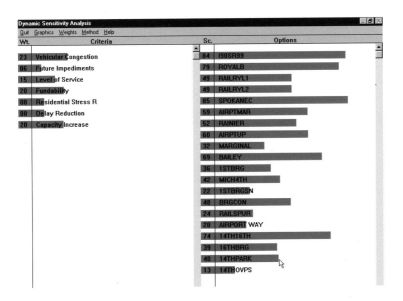

PLATE 10.3. Sensitivity evaluation of project ranking tied to a decision table in Figure 10.3. From Jankowski and Nyerges (2001a). Copyright 2001 by Taylor & Francis Ltd. Reprinted by permission.

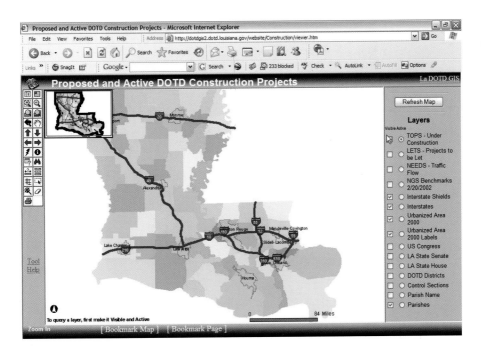

PLATE 11.1. Web viewer for proposed and active construction projects. From Louisiana Department of Transportation and Development (2006). Copyright 2006 by Louisiana Department of Transportation and Development. Reprinted by permission.

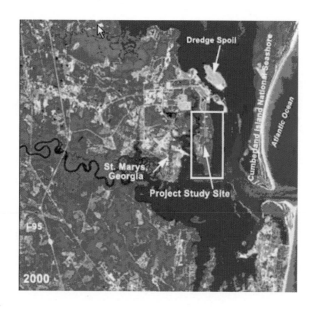

PLATE 11.2. Camden County, Georgia showing the town of St. Marys and the project site under consideration. From NOAA CSC (2006a). Reprinted by permission.

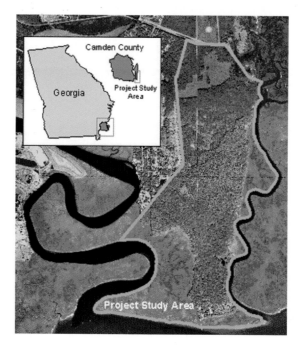

PLATE 11.3. The study site in relation to Camden County and the State of Georgia. From NOAA CSC (2006a). Reprinted by permission.

PART III

Planning, Programming, and Implementation Decision Cases

CHAPTER 8

GIS Data Analysis for Planning Decision Support

In several previous chapters we have made reference to planning activities. In this chapter we provide more detail about GIS-based data analysis for planning decision support. Planning is perhaps one of the most widely known, but little understood by the general public, processes carried out within communities among the three decision situation support contexts. One often-leveled criticism about planning is that never-implemented plans are not very useful plans. Thus, planning needs to be better connected with the other decision support situations, but we will get to that in the remaining chapters of the book. We start this chapter with an overview of planning analysis workflow in section 8.1. In section 8.2 we present comparative perspectives on planning-level analysis, characterizing land use, transportation, and water resources. We end the chapter with a summary.

8.1 Overview of Planning-Level Analysis Workflow

Plan making and plan use involve at least four major workflow components: standards of rationality, processes, tasks, and behaviors (Hopkins 2001). Those four components characterize the overall workflow construct of a decision situation framework presented in Chapter 4, focusing on plan making. Let us describe them from the most specific to the most general.

- *Planning behaviors*—fundamental actions that people take when they are making or using plans (talking to a constituent, involving participation of citizens, coloring a map, setting up date, etc.).
- *Planning tasks*—combinations of planning behaviors that accomplish certain functional purposes (e.g., forecasting, evaluating two options).

- *Planning processes*—patterns of tasks; planning processes yield plans.
- Standards of rationality—provision of criteria by which to judge planning processes; standards of rationality are different than a rational procedure itself; the standard is what sets the guideline for a procedure.

Organizational mandates, such as laws and regulations, motivate the making of plans—the move toward the action of plan making, with a plan as the outcome. Growth management laws commonly require a plan that comprises several elements, such as land use, transportation, and critical resources, but not how to compose elements. Laws do not specify the details of how plans are to be made, because too much detail would make it awkward to introduce new processes that could be more efficient, effective, and equitable. Thus, local organizations establish the tasks and processes that they believe are most appropriate. Organizations use standards of rationality in part encouraged by law, as well from the planning literature, and in part from their own experience of what has worked to establish guidelines for processes. However, laws do draw from standards of rationality for certain characteristics that should be part of the plan-making process. For example, over the past several decades, many laws passed by national, state, and local legislatures require public participation in planning processes to ensure democratic processes.

Planning processes take on specific meaning when set in the context of standards of rationality. The processes yield the plans; however, the standards set the expectations for the workflow. Thus, the plans are only as good as the standards of rationality; we address the importance of standards of rationality for establishing rational procedures in section 8.2. Planning tasks take on specific meaning when set within the context of processes. Again, however, the actual accomplishments emerge out of the mutual expectations of those involved. Planning behaviors take on specific meaning when set in the context of tasks to be accomplished; the actual accomplishment emerges from joint and/or conflicting expectations of people's behaviors during the process. This is a multilevel process problem; it follows the macro-phase and micro-activity approach presented in section 4.2.3 that helps us unpack and simplify the complexity of process.

Consequently, plan-making processes are commonly described by listing a sequence of tasks, or what we referred to as *phases* when we described GIS workflow tasks in Chapter 3. That sequence can be iterative; that is, a process is commonly a multipass activity. For a workflow of a plan-making process we once again turn to Carl Steinitz's work. Several GIS project activities undertaken by Steinitz and his colleagues show that it is possible to undertake procedural and communicative rationality at the same time. A widely recognized application of the landscape (broader than land use) planning process is the Camp Pendleton project (Steinitz 1996). Portions of Camp Pendleton, a military reservation north of San Diego, are reverting to civilian use. A long-term land use plan for the area was needed to provide guidance for what the alternative future uses might be. That project is one among several successful projects that use the same six-phase process to organize land use plan making (see Steinitz 1990, 1996; Steinitz et al.

2003). The Steinitz landscape planning workflow framework suggests using a three-pass process. Figure 8.1 depicts a top-to-bottom scoping pass, a bottom-to-top design pass, and top-to-bottom implementation pass for the overall process.

More recently, Steinitz and several colleagues (2003) used a similar approach in the Upper San Pedro River Basin in Arizona and Sonora Mexico. This GIS-supported project was an international dilemma in habitat preservation across the upper portion of the entire San Pedro River Basin, set in the context of a regional development plan. The six-phase landscape modeling process combined a procedural rationality with a communicative rationality to devise alternative futures for the landscape. This planning activity, once again, was considered a highly successful use of GIS (and, of course, other modeling software), providing a local community and the U.S. Army federal funding agency on the project with insights for how to manage a critical ecosystem of plants, animals, and humans.

8.2 Comparative Perspectives on Planning-Level Analysis

A planning data analysis takes a broad-based approach to data analysis (lots of area and considerable time horizon; e.g., 20 years) in comparison to other analysis levels. In a planning analysis, a depth of understanding at a particular site is not as important as understanding how project sites interrelate over time across a broad-based geographic domain. The entire planning area does not necessarily have to change, but portions of it do take on some change. It is true that we can speak of small-area analysis planning, but this type of analysis is a scaled down version of larger area planning analysis and can also include project implementation analysis in the mix. In growth management contexts, plans establish how projects together might transform a landscape. The projects and, therefore, the plans are often conceptual in nature, in that the details are not readily known. It takes time and resources to come to understand that detail, which is why planning differs from improvement programming, and improvement programming differs from project implementation.

8.2.1 Comparing Land Use Planning Analysis Workflow

Hopkins (2001) compares multiple planning process sequences to show that many of the descriptions are similar, and only some are contradictory. The processes are described at different levels of task resolution, ranging from three to eleven tasks, with some steps being more general than others. To make sense of the differences, Hopkins (2001 p. 192) compares procedural rationality and communicative rationality (see Table 8.1). He views "rationality" more thoughtfully as a standard of performance rather than as a process, whereby a different performance is achieved with procedural goals than with communicative goals. Procedural goals deal with analytic achievements to complete a set of steps, whereas communicative goals deal with deliberative achievements as in everyone's voice is heard with regard to their interests.

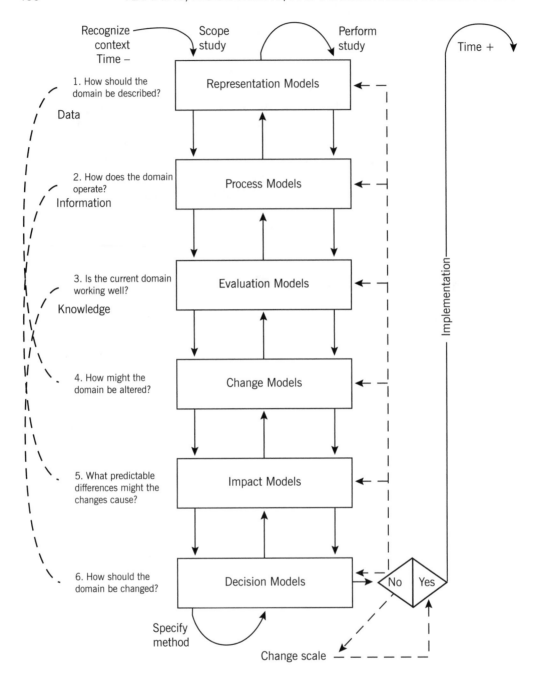

FIGURE 8.1. Workflow for analysis at the scale of regional planning. Adapted from Steinitz et al. 2005. Copyright 2005 by Carl Steinitz. Adapted by permission.

TABLE 8.1. Comparing Procedural and Communicative Rationality

Procedural rationality	Communicative rationality
• All goals considered	• All interests represented
• All aspects of current and future situations assessed	• Interests informed and able to converse about situation
• All alternatives considered	• Interests equally empowered
• All impacts from alternatives tested	• Good reasons, good argument
• All alternatives evaluated on all criteria	• Allow all claims and assumptions to be questioned
• Best alternative selected	• Consensus reached

Note. Adapted from Hopkins (2001). Copyright 2001 by Lewis D. Hopkins. Adapted by permission of Island Press, Washington, DC.

The comparison is not an either–or circumstance. Merging the two rationalities into a single overall process is appropriate, taking advantage of both goals, as in the analytic-deliberative process suggested by the National Research Council (1996) with regard to environmental decision making.

So what is the relationship between "goals" in procedural rationality versus "interests" in communicative rationality? Both are grounded in values, if we use a value tree (structure) perspective. *Goals* are general statements of outcomes or processes that should be addressed; a level of achievement to be reached. Interests in a decision problem are more deep seated with regard to contextualized concerns. Nevertheless, goals and interests can be shared as demonstrated by Kenney (1992) in his work on valued-based thinking. Discourse is important to having goals and interests in alignment.

Innes (1996, 1998) emphasizes direct implementation of communicative rationality as a decision process, which means there is a close relationship with procedural rationality. She deals with cases that commonly form ad hoc institutions to address a problem or issue. Thus, Innes believes it is possible to design rational institutions for particular situations. But such a design requires a discourse that delves deeply into what people value.

On the topic of values in planning, Hopkins (2001) states that ...

Connecting actions to consequences requires not only forecasting effects but also valuing those effects to determine preferences among options. Techniques for eliciting values and preferences independent of specific options—either alternative actions or alternative outcomes—are notoriously difficult to implement. Either people respond in ways that do not express their intended preferences or they refuse to respond at all (Lai and Hopkins 1989, 1995; Lindsey and Knapp 1999). The questions often used to elicit preferences in practice do not yield valid information, even assuming that people know what their preferences are. Valid methods are either too costly to implement for practical use, or people refuse to use them because the questions are simply too difficult to answer. Computing tools may be useful in reducing the number of such trade-off questions needed to make a choice (Lee and Hopkins 1995). (p. 205)

Sensitivity analysis (SA) methods (see section 7.4) help identify critical evaluation criteria, for which preferences must be known, as opposed to noncritical criteria, for which the knowledge of preferences is irrelevant for evaluating plan options. Knowledge of such criteria can be useful in reducing the number of trade-off questions needed to elicit preferences (Salteli, Chan, and Scott 2001).

According to Hopkins (2001 p. 218), there are several claims to consider when making plans, based on analytical and empirical evidence:

1. Recognize opportunities to use plans. Look at plans from the perspective of decisions about available actions (whether there is a plan that is appropriate for guiding decisions).
2. Create views of plans for decision situations. Do not present plans only from the perspective of plan making.
3. Recognize opportunities to make (revise) plans. Consider decision situations as an opportunity to revise a plan. Do not make plans at fixed time intervals for fixed time horizons (opportunities for change recognize leverage points; something must be controllable to be able to create a work program to specify a budget).
4. Make plans of efficient breadth by choosing the useful functional, spatial, and organizational scope of a plan in shaping and making decisions. Do not strive for a comprehensive plan on the premise the closer to this ideal, the better.
5. Link consequences to interdependent actions that are available, and recognize the uncertainty in these links and these actions.
6. Use formal and informal institutions to deliberate and act. Do not privilege direct participation as if it were inherently more effective or fair than routine process, and do not presume that plans are or must be collective choices.

Many of these opportunities and claims are actually part of practice. In light of these opportunities and claims, we present a sample description of land use planning as it is practiced in the state of California (California Resources Agency 2005). In this description we focus on the components of land use planning and on resulting spatial information needs.

Incorporated cities and counties in California are required by state law (planning, zoning, and development laws passed in 2000) to adopt a comprehensive, long-term general plan for their physical development. This general plan is the official city or county policy regarding the location of housing, business, industry, roads, parks, and other land uses. The legislative body of each city (the City Council) and each county (the Board of Supervisors) adopts zoning, subdivision, and other ordinances to regulate land uses and to carry out the policies of its general plan. Because cities and counties are treated as independent entities, there is no requirement that adjoining cities or counties have identical, or even similar, plans and ordinances. In most of the communities, the City Council or the Board of Supervisors has appointed one or more hearing bodies to assist them with planning activities. The common types of hearing bodies and their usual responsibilities include the *Planning Commission*, which considers general plan and specific plan amendments, zone changes, and major subdivisions; the *Zoning Adjustment Board*, which considers conditional use permits, variances, and other minor permits, and the *Architectural Review Board*, which reviews projects to ensure that they meet community aesthetic standards.

The local general plan representing the community's vision of the future is made up of goals and policies upon which the City Council, Board of Supervisors, or Planning Commission makes land use decisions. General plans comprise (1) a written text discussion of the community's goals, objectives, policies, and programs for the distribution of land use, and (2) one or more maps illustrating the general location of existing and future land uses.

The general plan is different from a zoning policy in that the former takes a long-term outlook, identifying the types of development that will be allowed, the spatial relationships among land uses, and the general pattern of future development. Zoning policies regulate present development through specific standards, such as lot size, building setback, and a list of allowable uses. The land uses shown on the general plan diagrams are usually reflected in the local zoning maps as well. Development must meet not only the specific requirements of the zoning ordinance but also the broader policies set forth in the local general plan.

Each local general plan in California must contain the following seven elements:

- The *land use element* designates the general location and intensity of housing, business, industry, open space, education, public buildings and grounds, waste disposal facilities, and other land uses.
- The *circulation element* identifies the general location and extent of existing and proposed major roads, transportation routes, terminals, and public utilities and facilities. It must be correlated with the land use element.
- The *housing element* is a comprehensive assessment of current and projected housing needs for all economic segments of the community. It sets forth local housing policies and programs to implement those policies.
- The *conservation element* addresses the conservation, development, and use of natural resources, including water, forests, soils, rivers, and mineral deposits.
- The *open space element* details plans and measures for preserving open space for natural resources, the managed production of resources, outdoor recreation, public health and safety, and the identification of agricultural land.
- The *noise element* identifies and appraises noise problems within the community and forms the basis for distributing new, noise-sensitive land uses.
- The *safety element* establishes policies and programs to protect the community from risks associated with seismic, geological, flood, and wildfire hazards.

At the same time, each jurisdiction is free to adopt a wide variety of additional elements, such as recreation, urban design, or public facilities. It is clear from this list that spatial information needs of local general planning are diverse and cut across land use, transportation, and water resource issues. In fact, the policies determining future land use patterns codetermine future circulation, housing, conservation, open space, noise, and safety conditions.

Spatial data needs of land use planning include soil surveys that classify, describe, and map soil properties (e.g., physical and chemical composition, erosion, fertility, permeability). In the United States, soil series maps in the digital format are produced by the Natural Resource Conservation Service (NRCS). Maps at the scale of 1:24,000 include over 50 soil properties, many of them useful in land use planning. Other spatial data needs for land use planning include vegetation maps derived from the interpretations of satellite imagery and aerial photographs, topography, and existing infrastructure maps, including, roads, sewer and other utility lines, and maps of ecologically sensitive areas.

Some of these datasets are represented by contiguous (raster) map layers (e.g., vegetation and topography) and others by feature-based (vector) maps layers. Land use planning also requires population data, which can be derived from the U.S. Census 2000 for various geographical subdivisions, ranging from individual city blocks to census block groups to census tracts.

8.2.2 Comparing Transportation Planning Analysis Workflow

Sixteen years after the publication of the first edition, Meyer and Miller (2001) revised their book *Urban Transportation Planning* with the intent to recharacterize the metropolitan transportation planning process from one that focused on technical analysis and transportation modeling in the early 1980s, to one that today puts decision-making processes at the center of the planning process.

> Because transportation is critical to the social, environmental, and economic health of every metropolitan area, decisions to change this system must be considered with knowledge of the likely impacts or proposed actions and of the consequences if no decision is made. The underlying premise of this book [urban transportation planning] is that, in order to provide such knowledge effectively, the planning process and the related technical analysis should be consistent with the substance and form of transportation decision making. There are however, many different ways of viewing the decision-making process. Understanding the nature of alternative decision-making processes and the needs and capabilities of those who participate in them are thus prerequisites for the development of an effective transportation planning process. (p. 41)

Given the many factors that influence how transportation decisions occur and why certain choices are made, there is no single general description of decision making that applies in detail to every situation. However, Meyer and Miller offer a common set of characteristics of all transportation decision making that occurs within an "institutional framework." They describe such a framework as having five components that behave and link in various ways within regions to make transportation systems what they are. First, *organizations* provide and manage transportation services, with each having specific (and sometimes competing) mandates. Second, the *formal process of interaction among, and production of outputs from, these organizations* is often mandated by other levels of government (as such, there is a governmental hierarchy often associated with scale of jurisdiction). Third, *informal personal and group dynamic relationships* make the process work (or slow it down, as Meyer and Miller observe). Fourth, the *political, legal, and fiscal constraints* (guidance in mandates) can either provide strong support for desired outcomes or become nearly insurmountable barriers with regard to certain values and objectives of various groups within the community. Fifth, there are *positive and negative roles of specific individuals or groups* associated with responsibilities mandated, granted, or adopted by these individuals and groups. The dynamic nature of the urban transportation system, and the institutional environment within which it operates, suggests that dealing with change will be a continuing characteristic of transportation planning.

Today, in a typical metropolitan area, a large number of organizations are involved with transportation planning and decision making. In the United States, federal regulations require each urbanized area with a population over 50,000 to have a metropolitan planning organization (MPO) that is responsible for transportation planning (at the regional level). An MPO has five core functions:

1. To establish and manage a level playing field for effective multimodal, intergovernmental decision making in the metropolitan area.
2. To develop, adopt, and update a long-range multimodal transportation performance plan for the metropolitan area that focuses on three types of performance: mobility and access for people and goods, system operation and preservation, and quality of life.
3. To develop and continuously pursue an appropriate analytical program to evaluate transportation alternatives and support metropolitan decision making, scaled to the size and complexity of the region, and to the nature of its transportation issues and the realistically available options.
4. To develop and systematically pursue a multifaceted implementation program designed to reach all metropolitan transportation plan goals, using a mix of spending, regulating, operating, management, and revenue enhancement tools.
5. To develop and pursue an inclusive and proactive public involvement program designed to give the general public and all the significantly affected subgroups access to and important roles in the four essential functions listed above (Advisory Commission on Intergovernmental Relations 1997).

Almost half of the MPOs in the United States are regional councils of government, with memberships that comprises mainly local elected officials, and are responsible for coordinating transportation under state mandates (Advisory Commission on Intergovernmental Relations 1997). In addition to the MPO, most regions have regional and local transit providers; county or local transportation or public works agencies; city planning departments; social service agencies providing transportations services; and special authorities, such as parking, airport, port, or recreational districts with transportation responsibilities.

> In addition to government agencies, participants in the planning process can include a variety of private sector organizations and community groups. In recent years, transportation legislation and regulations have provided for a much more inclusive process that promotes participation by anyone who wants to be involved. Typical participants include environmental groups, community associations, bicycle/walking advocates, business associations, civic groups, freight providers, groups focused on specific issues (e.g., freight mobility), government groups, developers, trade associations, representatives from disadvantaged populations, professional organizations, and tourism groups. (Meyer and Miller 2001 p. 45)

As one would expect with such a diverse set of actors involved with transportation, transportation planning and decision making are very complex. The MPO is responsible for coordinating the participation of all these groups in the planning process. Because so many stakeholders can be part of this process (with so many different interests), a challenge for transportation planning is to provide as much information as possible on the consequences of alternative decisions, so that decision makers understand who gains and who loses from the decision. In addition, with such a diverse set of interests potentially engaged in the process, transportation planning often becomes a means of educating groups on the underlying causes of transportation problems and on the likely results of actions to solve them.

That challenge, referred to earlier in the context of diverse interests, becomes one of framing in as broad a way as possible the transportation analysis that generates the options and at the same time informing the deliberation that addresses those options in such a way that option impacts are understood by all groups about who gains and who loses.

The U.S. Federal Transportation Law, the Transportation Equity Act for the 21st Century (1998) contained language whose intent was to streamline the environmental analysis and project development process. As stated in the proposed regulation to implement this requirement, "a new approach to NEPA [National Environmental Policy Act] is needed, one that emphasizes strong environmental policy, collaborative problem solving approaches involving all levels of government and the public early in the process, and integrated and streamlined coordination and decision making processes" (Meyer and Miller 2001 p. 46). Importantly, "the transportation planning process needs to be coordinated with the project development/NEPA process so that transportation planning decisions can alternately support the development of the individual projects which arise from the transportation plan" (Meyer and Miller 2001 p. 46). The major intent in this policy was to reduce the amount of time it takes for a project to be developed, but at the same time make environmental analysis more effective.

> Because transportation provides opportunities for social, economic, and community activities, policy makers have turned to the transportation sector as a means of achieving a vari-

ety of societal goals. In so doing, they have linked (but not often integrated) the transportation planning process to other planning and policy initiatives. Policy linkages exist between transportation and growth management, welfare, energy, and other environmental policies. The net effect on the institutional environment for transportation is once again to broaden the number and variety of actors involved in the planning process. (Meyer and Miller 2001 p. 46)

So how can members of an organization (and, for that matter, an interorganization partnership) go beyond the link and integrate the planning initiatives if they so desire? Is there an information tool within an information systems context that supports such integration? We propose that a participatory values articulation approach and values tree tool can connect a variety of values, objectives, and indicators of what is to be the focus, the objective, and the state of change to be brought about by transportation project/plan development.

The mix of strategies considered in the transportation planning process outlined by Meyer and Miller (2001) leads to greater participation in the planning process by a variety of groups.

> Given that the mix of strategies is so diverse, a large number of organizations, groups, and individuals will likely be involved in any major planning initiative in a region. This means that the planning process will have to provide participation opportunities for a diverse set of interests, each having different perspectives on the likely consequences of project/plan alternatives. (p. 48)

Meyer and Miller provide a perspective on the evolving nature of the metropolitan transportation planning and decision-making process. They describe a number of conceptual models for characterizing decision making, from which they extract several "elements" of decision making to provide an overview of a "decision-oriented planning process" . That overview includes the following characteristics for what a 21st-century transportation "decision-oriented planning process" should do: (1) establish a future context, (2) respond to different scales of analysis, (3) expand the scope of problem definition, (4) maintain flexibility in analysis, (5) provide feedback and continuity overtime, (6) relate planning to programming and budgeting process, and (7) provide opportunities for public involvement. With these characteristics in mind, they outline four major stages of a decision-oriented transportation planning approach:

1. *Problem identification and/or definition.* This is a matter of clarifying perceived differences in current and desired states of affairs and interpretations of situations.

2. *Debate and choice.* This involves making sure a set of feasible alternatives is part of the decision mix, recognizing limited resources, the need to set priorities, and the selection of one or more alternatives within an atmosphere of conflict due to differences in values, objectives, interests, and/or interpretations of data.

3. *Implementation.* Beyond the mere choice being made is the actual process of putting that choice into action, as in implementation. Implementation of plans through programming of projects is the linkage between planning and programming that is now being recognized as a gap in the process of how better to coordinate change in transportation systems.

4. *Evaluation and feedback.* Recent federal transportation laws have made it clearer that understanding transportation system performance is a matter of monitoring appropriate characteristics through performance measurement (Advisory Commission on Intergovernmental Relations 1997). Providing appropriate feedback in the short, medium, and long term can provide perspective about how well the decision process is addressing the perceived needs in problem identification/definition.

In empirical research about information technology development, what is a normative four stage process to Meyer and Miller (2001), might actually be seen by local participants as a three-, five-, six-, or seven-, and so forth, stage process, depending on how each of these phases is articulated given the place and time planning context within central Puget Sound. However, we should be aware that in a place and time context, whether fewer or more stages are needed, each of these concerns would need to be addressed. It is simply a matter of developing a shared understanding about the decision process among those responsible for this process. For example, Meyer and Miller show how the four-stage framework maps onto a generalization of an urban transportation planning process with 11 planning activities (Cambridge Systematics 1996). Thus, the multiple steps of planning relate to a single phase in the four-phase process. The 11-phase process was developed as a National Cooperative Highway Research report, and supposedly can apply to many metropolitan contexts around the United States (see Table 8.2).

What we can say is that the four-stage overall decision process is but an abstract of the 11 steps and suffices to "start a conversation" about how geospatial information technologies can support a decision process. One can start with the four phase process or the 11-step process and apply the decision situation assessment framework (Jankowski and Nyerges 2001a) to elucidate decision support needs in participatory settings. Each macro phase becomes the basis of a task in an information needs investigation directed at guiding decision support tools development to link long- and short-term planning (i.e., supporting the development of a decision-oriented urban transportation planning process).

It is indeed a complex endeavor but, again, the process is somewhat long term, with many people involved. Nonetheless, the process should be transparent to analysts, professional planners, and citizens alike if the plan is to receive support from a diverse community. Below we present an example of transportation planning as practiced in the central Puget Sound Region of Washington State. This example stems from a case study of transportation planning at the Puget Sound Regional Council. GIS provides a supporting rather than central role. The Emme/2 software used for transportation forecasting plays the central role in estimating traffic volumes by mode for all links of the transportation system (INRO 2008).

TABLE 8.2. Comparing Workflow Phases Presented by Meyer and Miller (2001), the National Cooperation Highway Research Program (NCHRP; Cambridge Systematics 1996), and Steinitz et al. (2003)

Meyer and Miller decision-oriented framework associated with planning process	NCHRP urban transportation planning process, adapted by Meyer and Miller (2001)	Steinitz landscape planning (modeling) framework
Problem identification/ problem definition	"Vision" expressed in terms of a triangle with nodes labeled • Prosperity • Quality of life • Environmental quality Each is related to the others.	Representation modeling
Problem identification/ problem definition	Goals and objectives	Representation modeling
Evaluation and feedback	Performance measures	Representation modeling
Debate and choice	Data	Representation modeling
Debate and choice	Analytical methods	Process modeling Change modeling Impact modeling
Debate and choice	Alternative improvement strategies	Decision modeling
Debate and choice	Other sources for project ideas	Representation modeling
Debate and choice	Evaluation criteria	Evaluation modeling
Implementation	Fiscal and resource prioritization	Decision modeling
Implementation	Implementation of strategies	
Evaluation and feedback	System operations	Evaluation modeling

The analysis context is portrayed in Figure 8.2 (p. 176). One can see that transportation analysis is at the core of the content but not the entire context. The analysis steps are outlined in Figure 8.3 (p. 177). The process for modeling comprising 10 steps is rather complex.

The analysis steps are performed using the Emme/2 (and now Emme/3) forecasting software (INRO 2008). Because Emme/2 software was developed specifically to perform travel demand forecasting, it does not perform many of the functions that GIS software performs. Consequently, each supports the other in the transportation planning process. A GIS data management environment often supports the development of the transportation network. Although GIS plays a supplemental role in the analysis, it plays the central role in data management of the network, as well as in visualization of the inputs, manipulation, and outputs (see Figure 8.4, p. 178).

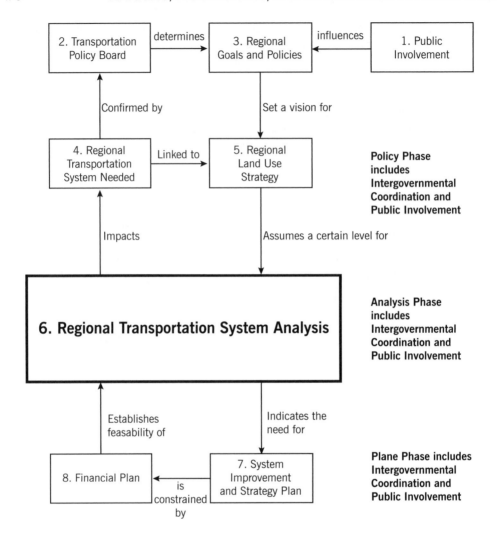

FIGURE 8.2. Transportation plan-making process of the Puget Sound Regional Council circa 1995. Adapted from Nyerges (1995). Copyright 1995 by The Guilford Press. Adapted by permission.

8.2.3 Comparing Water Resource Planning Analysis Workflow

Dzurik (2003) describes water resource planning as being comparable to other types of planning. He characterizes the nine-step planning process as follows:

1. Problem identification
2. Data collection and analysis
3. Development of goals and objectives
4. Clarification and diagnosis of the problem or issues
5. Formulation of alternative solutions
6. Analysis of alternatives

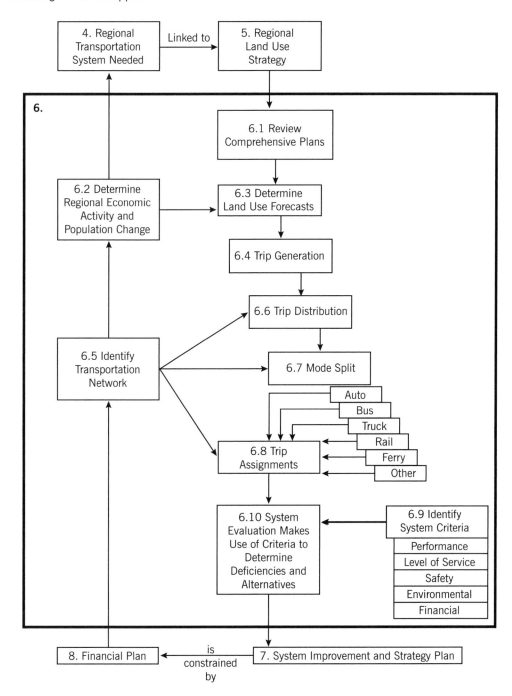

FIGURE 8.3. Steps in transportation (travel demand forecasting) analysis process. Adapted from Nyerges (1995). Copyright 1995 by The Guilford Press. Adapted by permission.

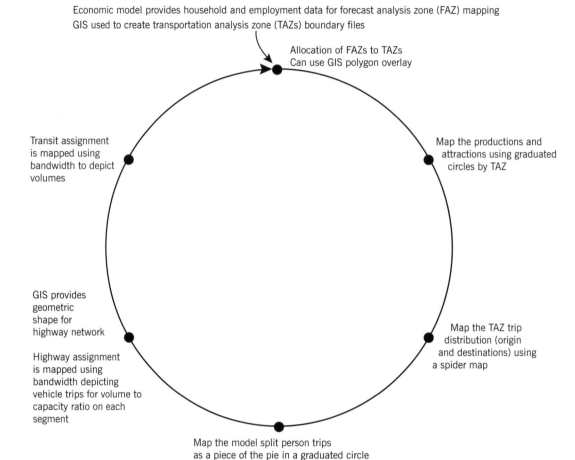

FIGURE 8.4. GIS support for travel demand forecasting. From Nyerges (1995). Copyright 1995 by The Guilford Press. Reprinted by permission.

7. Evaluation and recommendation of actions
8. Development of an implementation program
9. Surveillance and monitoring

We elaborate on each of these steps as a workflow list of tasks, as follows:

1. *Identify the problem*
   a. Identify needs and concerns with respect to the water resources of an area, whether local, regional, or national. Needs for planning can stem from:
      • Experiencing a problem, such as flooding, or an anticipated problem, such as inadequate water supply to meet future needs.
      • Generalized concern and not dealing with any single problem or need (e.g., river basin studies and areawide water quality studies).

   b. Identify and clarify competing and conflicting interests involved.

   c. Involve public and begin/further coordination with agencies and groups.

2. *Collect and analyze data from available data stores.*

   a. Define study area (e.g., catchment, subwatershed, watershed, basin).

   b. Identify existing data pertinent to the problem—for example, geophysical and biological features; social, demographic, and cultural characteristics; and land uses and economic activity (e.g., manufacturing or commercial). Identify the current data systems at the federal and regional level, for example, Hydrological Benchmark Network, National Stream Quality Accounting Network, Water Data Storage and Retrieval System, and Storage and Retrieval System.

3. *Develop goals and objectives.*

   a. Specify relevant goals from organizations, the public, and groups involved. Planning goals are broad and general, and relate human values to natural resources and environment (e.g., attain clean water or provide adequate water supplies). These can be derived from agency and unit mission or policy statements, and the laws that encourage these agencies to act/protect in a particular way. Implementation goals are specific, for example, a level of clean or what it means to be adequate; such a goal could be an extant standard for water quality.

   b. *Identify objectives.* Objectives are more specific statements associated with planning goals, for example:

     • Prevent continued water degradation by waterborne wastes.

     • Prevent or reduce flood damage.

     • Provide recreation.

     • Provide efficient reuse of treated wastewater.

     • Provide efficient development of water supplies.

Objectives indicate an interest in addressing a dimension but do not indicate to what extent these should be addressed.

4. *Clarify and diagnosis problem or issues.* Use the data previously identified or collect new data as directed, in combination with the goals and objectives to characterize the problem condition in terms of those goals and objectives. What is the current state of the water resource conditions, as well as social, economic, and environmental conditions associated with the problem? Sometimes, new data need to be collected to address certain objectives for which data do not currently exist. A reformulated (updated) understanding of the problem provides a basis for developing alternative solutions. Compare the difference between the decision goal and the extant condition for each objective. This difference provides the basis of the problem gap and suggests the dimensionality of solutions sought.

5. *Formulate alternative solutions.* An analyst can use the objectives and data previously articulated to establish criteria measures to formulate alternatives; alternatives should be as narrow as necessary to address the objectives; alternatives should be suf-

ficiently broad to address all objectives; and/or alternatives should be formulated, with decision goals compared to measured conditions to close the gaps.

6. *Analyze alternatives.*

    a. Analyze the aggregate problem gap closure by each alternative solution (plan). Which ones address the problem best in regard to social, economic, and environmental concerns? A cost–benefit analysis can be used for economic evaluation; multiattributed utility theory can be used to enumerate social and environmental measures.

    b. Perform an impact assessment by enumerating the impact of that gap closure. What are the implications of closing the gap overall for each of the alternative solutions?

7. *Evaluate and recommend actions.*

    a. Identify the issue and objective to which each alternative is directed.

    b. Determine the positive and negative character of the alternatives, using both public input and professional evaluation.

    c. Display the results of the evaluation, so that decision makers know how each alternative relates to local, regional, state, and national issues and policies. This display includes the trade-offs for the alternative (i.e., the results of gap analysis for each alternative).

    d. Consider the following:

- Determine how well alternatives satisfy decision goals.
- Compare the aggregate results of that satisfaction, one alternative to another.
- Analyze the trade-offs (the differences among those gaps; e.g., economic efficiency and environmental quality).

Criteria suggested by the US Army Corps of Engineers

- Acceptability—Is the plan workable and viable for affected parties and known institutional constraints?
- Effectiveness—Rate the plan's technical performance and contribution to planning objectives.
- Efficiency—Can the plan meet objectives functionally in least costly way?
- Completeness—Are all necessary investments to attain the plan included?
- Certainty—What is the likelihood of the plan to meet planning objectives?
- Geographic scope—Is study area large enough to address the problem fully?
- Cost–benefit ratio—Determine economic effectiveness in relation to planning objectives.
- Reversibility—Measure the capability to restore a complete project to original condition.
- Stability—Analyze sensitivity of the plan to potential future developments.

8. *Develop an implementation program.* At this point, the plan is adopted and put forward to guide programs in which projects are selected for design and construction. Often, in water resources planning, considerable effort goes into developing plans that

are never adopted. In some cases, a plan is approved and adopted but is never carried out, or it is set aside for years. Why is this often the case? Dzurik (2003) never addressed the issue. Is it because things change? But plans are meant to guide. Perhaps institutional memory over the short term is sufficiently "steeped" in the process, such that the organization has learned something new. Perhaps different parts of an organization are responsible for planning, programming, and implementation. Meyer and Mitchell (2001) address this same issue in transportation, and hint that different parts of an organization may find it challenging to communicate. In addition, we contend that databases developed as separate representations of the world encourage people who are responsible for those databases to view the world in a disjointed manner. This suggests that disconnects between planning and improvement programming, and project implementation are due in part to a lack of data integration. We return to that subject in Chapters 12 and 13.

9. *Perform surveillance and monitoring.* Plan implementation should be compared to the original goals and objectives of the plan. Because many plans take so long to implement, conditions change, and because needs change, the implementation is different than expected. In such a case, the plan should be updated to reflect those needs.

State-level comprehensive planning initiated in the 1960s and 1970s has been replaced with focused executive planning and policy in states. Comprehensive planning at the local level in the 1990s was instituted to address many concerns about the impacts of rapid population and economic growth on environmental conditions. The trend in planning at the state level now recognizes limitations in an ability to be truly comprehensive given the especially fast-changing policy and decision-making environments. Long-range planning now relies more on incremental learning and feedback mechanisms as a way to deal with complexity and change. Data management of conditions has certainly been challenged.

Another approach to the water resource planning process is presented by Beecher and Shanaghan (2002), who suggest that several key questions addressed during each step of a water resource planning process (see Table 8.3).

Several of these questions beg answers from GIS data processing. Here we highlight only a few, because we turn to a case study about water resource planning in the next section. In Table 8.3 (step 3), the factors affecting water systems are often geographically dispersed. For example, identifying the consumers of water (e.g., residential, commercial, and industrial users) and the expected growth in that use provides the fundamental drivers. However, supply constraints (choke points) in the system can provide a clearer picture of the influence of drivers. In Table 8.3 (step 4), we want to know whether the system can respond to challenges and opportunities. How would natural hazards present challenges to a system? For example, would soil erosion of steep slopes present a hazard to piping systems? What about regional climate change and the availability of water? In Table 8.3 (step 5), we could identify several geographically distributed options to upgrade a water supply system. For example, where might groundwater and surface water influence each other, so that we should pay more attention to how they are interre-

TABLE 8.3. Planning Steps and Key Questions in a Small-System Context

Planning step	Key questions to guide planners
Step 1. Specify mission and goals	• What is the water system's mission? • What values guide the water system? • What are the system's immediate and long-term goals? • Are values and goals established in an open and participatory process that includes employees, customers, and other stakeholders?
Step 2. Assess structure and roles	• How is the water system structured in terms of the hierarchy from governance, to management, to operations? • What service functions does the water system presently provide? • What operational tasks does the water system perform? • What is the role of the water system for each service function?
Step 3. Identify challenges and opportunities	• What are the principal change factors or drivers affecting the water system? • What challenges are presented? • What opportunities are presented?
Step 4. Evaluate system capacity	• Does the water system have adequate technical, financial, and managerial capacity? • What are the system's strengths (performance-enhancing factors) and weakness (performance-limiting factors)? • Can the system manage change and effectively respond to external challenges and opportunities?
Step 5. Identify strategic options	• What strategic options are available to the system for achieving its goals? • What benefits and costs are associated with each option? • How are the system's technological and structural options interrelated?
Step 6. Choose a strategy	• Which strategic option (or combination of options) can best provide the system's service roles and functions? • How do options compare in terms of cost-effectiveness? • Which alternative is optimal in terms of the selection criteria?
Step 7. Implement	• What implementation issues are presented by the strategy, and how are they being addressed? • What will be the character of the improvement programming process to provide incremental financial build out of the plan?
Step 8. Monitor	• How will the strategy be monitored over time to ensure success? • Is the plan producing desired outputs and achieving desired outcomes?

Note. Adapted from Beecher and Shanaghan (2002). Copyright 2002 by Larry W. Mays. Adapted by permission.

lated. In the western United States there is considerable concern about the increasingly short supply of surface water. Having options to draw from groundwater, and where exactly to obtain the water, are two important considerations. We address this issue later in subsections in the planning case studies.

Yet another way to compare processes and, more importantly, the geospatial information needs to support these decision processes, is to examine processes side by side for two different organizations. Below we examine the statewide water resource planning processes for Texas and Wyoming, and assess how geospatial information is either different or the same providing the reader with a sense of how GIS workflow might change from one place to another.

8.2.3.1 Water Resource Planning in Texas

Motivation for a comprehensive statewide water plan in Texas, as in other semiarid states, derives from the awareness of the vulnerability to drought and the limits of existing water supplies to meet increasing demands as the population grows. The Texas population is expected to grow from about 19 million (in 2003) to more than 39 million people by the year 2050. Water planning in Texas is a bottom-up process, in which regional water planning groups are asked to prepare regional water plans for their respective areas. The regional plans are submitted every 5 years to the Texas Water Development Board, which approves and incorporates them into a 5-year, statewide water plan (Texas Water Development Board 2004).

Regional water planning groups are responsible for preparing and adopting regional water plans for their areas. Planning activities include developing the engineering, socioeconomic, hydrological, environmental, legal, and institutional components of the regional water plans. The planning process must provide for public input, hold public meetings, and furnish a draft report of the plan for public review and comments. Texas water laws require that each regional water plan address the needs of all water users and suppliers, except for certain political subdivisions that decide not to participate. Funding for the planning process is appropriated by the Texas legislature.

The planning process in Texas is focused on estimating water demands and supplies. Regional water planning groups are responsible for deciding how future water needs in their respective region may be met. Each regional water plan includes information about water supplies and demand, water quality problems affecting water supply, and social and economic characteristics of the region. The plan also identifies water supply threats to agriculture and natural resources. The following information needs are common to each regional water plan:

- *Information about water demand*: How much water is needed for domestic and residential, industrial, agricultural, and energy production uses?
- *Information about water supply*: How much water is available in each region during nondrought and drought periods? Where and when is there a surplus of supply or a need for additional supplies?

- *Information about strategies for meeting future needs*: What strategies meet future near-term needs (less than 30 years)? What specific options meet long-term future needs (30–50 years)? What are social and economic impacts of not meeting needs? Are there needs that cannot be addressed for lack of feasible solutions?
- *Information about ecologically sensitive areas*: Where are ecologically unique streams and rivers?
- *Information about water laws and regulations*: What regulatory, administrative or legislative recommendations can improve water resource management in the state?

8.2.3.2 Water Resource Planning in Wyoming

Due to topography and climate, Wyoming has been richer in water than Texas, with much of its water coming from annual snow pack. With a population of 493,782 (in 2000), which is only 2.4% of the Texas population and a much smaller future population growth estimate, Wyoming decided in 1997 that, after 24 years, it was time to update its 1973 framework water plan. The new planning process was based on two principles (Wyoming State Water Plan 2004):

- The process must originate from water users at the local level. Basin advisory groups of local citizens representing all water user interests are to be established for advising the Wyoming Water Development Commission and the State Engineering Office in both local basin issues and the public participation process for water plan implementation.
- The planning process must be based on a resource database that is accessible, easily updated as conditions change, and capable of providing accurate and timely information needed to make good resource management decisions at every level.

Wyoming's equivalent of the Texas regional water plans are water basin plans. There are seven major basins in Wyoming. According to the Wyoming Water Development Association, a voluntary organization of private citizens, elected officials, and representatives of businesses, government agencies, industry, and water user groups, water basin plan information needs include the following:

- *Information about water demand*: How much water is needed for domestic and residential, industrial, agricultural, and energy production uses? What are future potential uses of water resources for single purposes and conjunctive uses for economic growth, the environment, and aesthetics?
- *Information about water supply*: How much water is available from surface water, groundwater, snowpack, and annual precipitation?
- *Information about water transferability*: What is the feasibility of interbasin transfers of water?

- *Information about water laws and regulations*: Which state and federal laws, rules, and regulations are related to water resource development, use, and management?

- *Information about water resources infrastructure*: How is water delivered to present users and governing units? Does the necessary infrastructure exist for the development of unused waters, or for more efficient management of present water uses?

- *Information about cultural and social factors associated with water resources*: What cultural–traditional–heritage factors are associated with existing and historical water development, use, and management?

Comparing the information needs of water planning in Texas and Wyoming (Table 8.4), one can see that despite the differences, the information needs in Texas and Wyoming are similar. Water planning in both states requires information about the quantity of water supply and water demand disaggregated by management units, be it regional water districts or river basins. Information about the quantity of water supply can be derived from hydrological and hydrographical data, much of which is geographical in nature and can be grouped into geographical datasets. Hydrological data are concerned with describing the properties of movement of water and include stream flow characteristics, surface terrain, and rainfall response as a function of soil, vegetation, and land use. Hydrographic data describes water bodies, such as streams, rivers, lakes, and seas. Both categories of data, hydrological and hydrographical, including surface terrain data, are widely available for most of the water basins in the United States in digital formats compatible with major GIS software packages. Information about the quantity of water demand can be derived from land use and population data. Land use data in the digital format, derived from satellite imagery and aerial photography, have become available in recent years for most of the United States. With the arrival of the U.S. Census 2000, population data for the entire United States can now be obtained in the digital, GIS-

TABLE 8.4. Comparison of Water Planning Information Needs for Texas and Wyoming

Texas	Wyoming
Information about water demand	Information about water demand
Information about water supply	Information about water supply
Information about strategies for meeting future needs	Information about water transferability
Information about ecologically sensitive areas	Information about water resources infrastructure
Information about water laws and regulations	Information about water laws and regulations
	Information about cultural and social factors associated with water resources

compatible format for individual city blocks in urbanized areas and for census block groups in rural areas.

Additional water planning information needs include water quality and ecological characteristics of water resources, water transport and storage infrastructure, and water laws and regulations. Some data about water quality, ecology, transport, and storage infrastructure can be derived from the National Hydrography Dataset (NHD) developed by the USGS and the EPA. The NHD is provided for the entire country at the scale 1:100,000. Data about water quality are represented at monitoring points along the network. The EPA manages water quality by setting total maximum daily loads (TMDLs) for specific water bodies, such as stream segments or lakes. Data about transport and storage infrastructure facilities, such as dams, weirs, lock chambers, rapids, and bridges, are also represented by points or landmark features. In addition to NHD data about water quality, data about ecology and water system infrastructure are collected and maintained by states, local governments, and nongovernmental organizations (NGOs). For example, the city and the County of San Diego maintain a network of water quality monitoring points in a GIS-compatible format, and a local NGO—the Bay Keeper— conducts regular water quality monitoring at established monitoring points on streams discharging into San Diego Bay.

These comparisons show that there are several approaches, whether one is planning for land, transportation, or water resources. There are many similarities as well when we compare processes.

8.3 Summary

Plan making and use, whether land use, transportation, and/or water resources, have at least four major components involved in workflow. Planning behaviors are fundamental actions that people take when making or using plans (e.g., talking to a constituent, involving participation of citizens, coloring a map, setting up date). Planning tasks comprise combinations of planning behaviors that accomplish certain functional purposes (e.g., forecasting, evaluating multiple options). Planning processes are patterns of tasks, and it is these processes that yield plans. Standards of rationality provide criteria guidelines by which to judge planning processes, but these standards differ from the rational procedures that are the processes.

Hopkins has compared multiple planning process sequences to show that many of the descriptions are similar, and only some are contradictory with regard to ordering of tasks. The processes are described at different levels of task resolution. Because some steps are more general that others, there are different numbers of tasks. Furthermore, some do not exist in comparison to others. The behaviors, tasks, and processes are implemented in terms of standards of rationality, that is, an ideal description of what should be done. Procedural rationality and communicative rationality are but two of many standards presented and compared. The comparison is not an either–or circumstance. Merging the two processes is appropriate to take advantage of aspects of both rationalities. Procedural rationality is pursued in long-term plan making, because the process lays out a systematic and comprehensive set of steps, so that an organization(s) considers circumstances that are commonly unknown. In communicative rationality, a mismatch of interests is articulated as the basis of the problem. The combination of the two is very similar to the analytical/deliberative process recommended for complex environmental decision-making processes when the risks are high and the number of stakeholder groups is large.

No single general description of decision making applies in detail to all transportation planning situations, just as in the case of land use planning. We have described a framework with five components—organizations; formal processes of interaction; informal and group dynamic relationships; political, legal, and fiscal constraints; and positive and negative factors—that behave and link in various ways within regions to make transportation systems what they are. A GIS analyst's main concern for decision support is to understand how GIS can assist with those five components, paying particular attention to how the technology might challenge the behavior and linkage of those components. A transportation modeling workflow process devised by the Puget Sound Regional Council was presented as an example of how GIS can be incorporated into the process. A GIS data management environment often supports the development of the transportation network. Although GIS plays a supplemental role in the analysis, it plays the central role in data management of the network, as well as in visualization of the inputs, manipulation, and outputs, making it increasingly important to the planning process.

We compared water resource planning to other types of planning. Dzurik (2003) characterizes the planning process as nine steps: problem identification, data collection and analysis, development of goals and objectives, clarification and diagnosis of the problem or issues, formulation of alternative solutions, analysis of alternatives, evaluation and recommendation of actions, development of an implementation program, and surveillance and monitoring. An alternative planning process that includes a set of critical questions to be addressed is the basis of identifying several steps in which GIS can play a critical role. From there we compared the information needs of two water planning processes at the state level, those of Texas and Wyoming. Chapter 9 presents a case study from the Boise River Basin of Idaho to help us gain an understanding about how we might undertake planning by using a spatial–temporal approach.

8.4 Review Questions

1. Describe the relationship among planning behaviors, tasks, processes, and rationality. How do those concepts relate to the "logic of plans"?

2. What constitutes a workflow task model for planning-level analysis?

3. Describe the relationship between procedural rationality and communicative rationality. In what ways do they conflict? In what ways do they support each other?

4. How do planning-level analysis processes compare and contrast with one another? Is there a difference among land resource, transportation, and water resources?

5. What function do MPOs have within the United States?

6. Describe the potential for using GIS in a transportation planning process, as related to Figure 8.4. How does GIS software deal with network modeling for transportation systems?

7. What constitutes the general steps in a water resource planning process?

8. Why does it make sense that there is more than one way to perform planning for water resources?

9. How similar or different are the Wyoming and Texas processes for water resource planning?

10. What constitutes a workflow task model for water resources planning-level analysis for water resource management?

CHAPTER 9

A Case Study in Water Resource Planning Decision Support

In this chapter, we provide additional detail about GIS-based data analysis for planning decision support. A case study of the Boise River Basin regional water resource planning effort, convened by the Idaho Department of Water Resources, provides an example that incorporates both space and time into the planning process. First, we present background materials to set the stage for the decision problem. Then, we turn to a description of GIS-based operations used in the planning decision problem. We end the chapter with a summary.

9.1 Background on a Water Resource Planning Decision Problem

Water supplies in many areas of the arid western United States are inadequate to meet all demands. In the State of Idaho, and in all other western states, many streams and aquifers are unable to provide sustained water supplies that fully satisfy all uses during good water years, let alone drought years. The appropriation doctrine of "first in time is first in right" has provided a consistent basis for distributing limited supplies of surface water in Idaho. However, the impacts of groundwater pumping on surface water supplies have often been ignored because of the legal and technical difficulties that they invoke. With increases in groundwater diversions within the Boise River Basin, water deliveries must consider *conjunctive impacts*—interactions between groundwater and surface water—if fair delivery is to be achieved. An area map of a section of the Boise River Basin is shown in Plate 9.1. The Boise River Water Plan provides a 10-year conjunctive administration process for addressing the interaction between groundwater wells and surface water extraction. The focus of the decision problem is to develop a plan to manage a combination of groundwater wells conjunctively with surface water over a 10-year period. Consequently, the temporal aspect of the decision problem is as important as the spatial aspect of the problem.

A decision situation assessment case study was developed as a Boise River collaborative planning process, and we focus here on the planning process. The core of that process involves conjunctive administration. A conjunctive administration framework is outlined in Figure 9.1. In the context of a stakeholder group discussion, the foundational elements form the basis of a conjunction administration platform from which implementation recommendations are made. Part of that conjunctive administration platform is a GIS-based decision support software system called WaterGroup, which was developed with the ArcObjects software library from ESRI (Jankowski, Robischon, Tuthill, Nyerges, and Ramsey 2006).

Representation of data via GIS is fundamental to stakeholders' understanding of the technical information. Historical "black boxes" of information are opened and to a large extent can be depicted in an understandable manner with GIS displays. Using

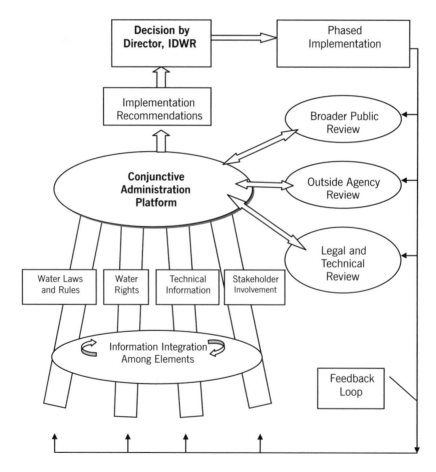

FIGURE 9.1. Conjunctive administration platform, including a GIS-based decision support software package called WaterGroup.

WaterGroup GIS, lay stakeholders are empowered to make recommendations given the data that are presented. Recommendations are subjected to broader public review, outside-agency review, and legal and technical review. The appropriate decision maker (Director of the Idaho Department of Water Resources) considers the recommendations and makes an initial determination. Additional review is conducted, and this process is repeated as additional information is developed in the basin of interest. Thus, stakeholders consider uncertainties in an iterative fashion *and become part of the decision-making process before the issue becomes unmanageable or contentious.*

9.2 Workflow Task Model to Guide the Analysis

Face-to-face decision sessions were conducted to create conjunctive administration scenarios. Each scenario proposed a 10-year plan, over the course of which groundwater wells would be administered together with surface water extraction from the Boise River. In one session, WaterGroup software was used by a facilitator–chauffeur (Plate 9.2), and displays were depicted on a single "public" screen to evoke discussion about water planning.

In a second session, several stakeholders used the software to explore the water planning options (See Plate 9.3).

The main toolbar in WaterGroup GIS comprises 10 tools (see Plate 9.4), organized into four categories (Gather, Organize, Select, and Review) that generally follow Simon's (1977) four phases of decision making (intelligence, design, choice, and reflect). This implementation demonstrates the opportunity to describe decision workflow using a macro-step and micro-activity framework for software design introduced by Jankowski and Nyerges (2001a).

The 10 tools on the WaterGroup toolbar are sequenced in order of anticipated use, with a first pass through the software. As such, the toolbar presents an expected workflow but does not dictate the workflow, because a user can make use of any tool at any time. The first tool provides access to "images" (see Plate 9.5). The red color shows vegetation, including riparian vegetation in which moisture is abundant, because this site is along the Boise River (near the top of the photo). Controls for working with imagery are in the upper left corner of the screen.

Two-dimensional displays are also part of the WaterGroup capability. The blue line in Plate 9.6 through the center of the map is the Boise River. Across the basin the hundreds of shallow wells (yellow color) are more numerous than deep wells (black color). The main aspect of the decision problem is to decide which wells, at what depth, are to be managed, starting in which year of the 10-year process. The wells shown in Plate 9.7 are a "representative" sample (928 wells) of all of the wells. There are literally thousands of wells, but a representative sample can be used to construct a plan.

Maps that show "flow detail" in terms of the response functions at particular locations are useful to indicate how the surface water flow in the river behaves given certain management scenario assumptions (see Plate 9.8). The *response function* (i.e., water drawdown at the upper and the lower ends of the basin section) are shown.

The well and flow details capability provides various charts in quadrants, with thumbnail windows in each of the display quadrants (see Figure 9.2). The details provide information on water quantity, viewed from various perspectives. Among the details are the well counts based on primary use of the wells over the 10 years, and the flow rate of wells based on primary use (upper left quadrant of Figure 9.1). *Primary use* would be residential, agricultural, industrial, and so forth. Another display is the well count by response function (i.e., how many wells are flowing at what rate when managed as part of the collection; upper right quadrant of Figure 9.2). The well count by the de minimis/ non–de minimis status of the wells is shown in the lower left quadrant of Figure 9.2). A de minimis status would belong to the smaller wells (commonly residential use), and the non–de minimis would be the larger wells (commonly agriculture or industrial). The response function by de minimis and non–de minimis status is shown in the lower right quadrant of Figure 9.2).

After a review of well locational distribution and depth displays, then an examination of some of the details regarding the situation, the capability to "define options" allows a user to establish a plan (See Figure 9.3). Each plan option is, in a sense, based on some assumptions, which is why we refer to these plan options as following a *policy scenario*. The plan is devised by setting wells at a specified distance from the Boise River and at a specified depth, to be managed in each of the years, starting with whatever year appears appropriate. However, all wells must be managed by the end of the 10-year period. Because the stakeholder groups have considerable knowledge about the area and the water demand for various groups, several plan scenarios were established from which to construct example plans.

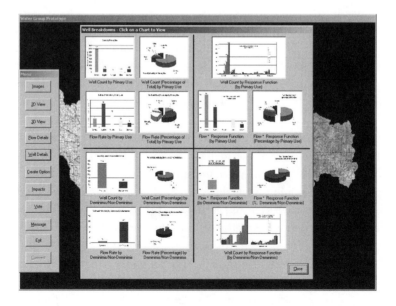

FIGURE 9.2. Flow detail shown as thumbnail windows in each of four quadrants.

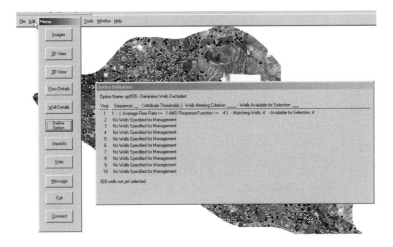

FIGURE 9.3. Defining options for a water plan make use of a query that specifies wells at a certain distance from the Boise River, and at a certain depth to be managed in each of the 10 years.

Once plan options are defined, the impacts of those options can be computed (Figure 9.4). The impacts are the aggregate flow rate and response function from all wells that are part of the plan. The objective is to manage all wells without adversely affecting any particular collections of wells to a significant degree in any given managed year.

Once the plan options are created, then the stakeholders can provide feedback on which plan they prefer or, in essence, they prioritize the plans given the circumstances at hand. The feedback comes in the form of a "vote" on the plans, so that a ranking of the

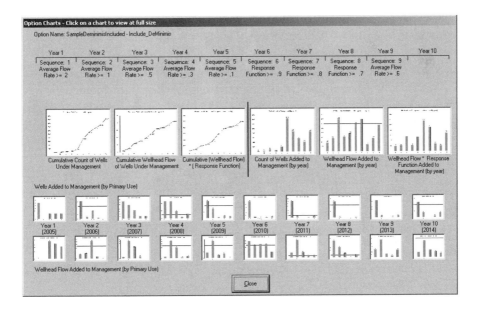

FIGURE 9.4. Impacts of the plan options.

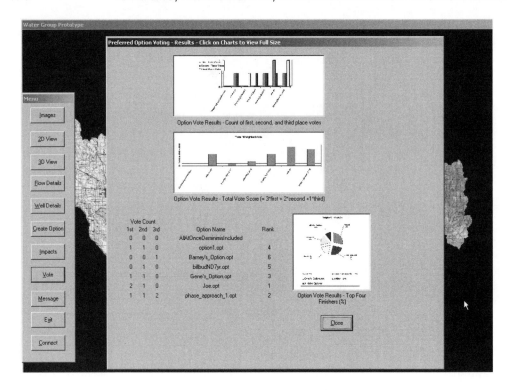

FIGURE 9.5. Voting results on plans.

plans can be established. Each stakeholder votes on the top three plans he or she prefers. The result is a ranked (i.e., prioritized) list of plans based on those votes (Figure 9.5).

This workflow process can be repeated as many times as necessary, until the stakeholder groups are satisfied with the results. Each pass takes about 3 hours to arrive at a ranked list of plans for a group of 12 people. After each planning meeting, the resulting plan rankings are forwarded to the Director of the Water Resources Department as a recommendation of how conjunctive administration might be implemented. The recommendation is not binding, because the Director is actually empowered by the legislature to make the decision. Each year such a plan ranking can be established, and forwarded, to adjust the direction of the plan. Because each plan incorporates a 10-year horizon, the plan years roll forward to reveal what has happened up to that time.

9.3 Summary

We have presented a case study about water resource planning that focuses on conjunctive administration of surface water and groundwater in the Boise River Basin of southwestern Idaho. The water plan outlines a conjunctive administration process for addressing the interaction between groundwater wells and surface water extraction. The focus of the decision problem is to develop a plan whereby a combination of groundwater wells are managed conjunctively with surface water over a 10-year period. The temporal aspect of the decision problem is as important as the spatial aspect. For that reason, specialized software is needed, in addition to basic GIS software, to address the planning problem. In general, we find that most organizations make use of several software packages to address all planning tasks. No single software can really do it all, but a software ensemble can readily get the job done.

9.4 Review Questions

1. What is conjunctive administration of water resources in the State of Idaho?

2. Why is it crucial to consider both space and time when making a conjunctive water resource plan?

3. What constitutes a workflow task model for a water resources planning-level analysis for conjunctive water resource management?

4. Why is specialized software needed, in addition to basic GIS software, to address most planning problems?

5. What additional benefit does highly interactive software provide when composing a plan?

6. What are the advantages and disadvantages of imagery versus maps, for example, the imagery depicted in Plate 9.5 compared to the well maps depicted in Plate 9.6?

7. Maps that show "flow detail" depict response functions at particular locations. What is the purpose of such maps?

8. Each alternative plan may be considered a space–time option. How were space–time options defined as part of the decision process?

9. What is the purpose of devising impacts?

10. How is voting used in the decision workflow?

CHAPTER 10

GIS Data Analysis for Improvement Programming Decision Support

GIS for improvement programming decision support is the most obscure of the three decision contexts among planning, improvement programming, and project implementation, because it is the least developed, hence the least utilized. Nonetheless, the improvement programming context has tremendous potential. The main reason is that improvement programming is the context in which the "budgeting of money" applies. Almost everyone is interested in setting priorities for budgeting money, as almost all adults do it, except, of course, for those who are unable to for one reason or another. The reader must understand that much of the world has yet to discover the capabilities of GIS. Some applications come before others in many respects, because software has been made available to do it. This has not really been the case for improvement programming. Why? One of the challenges with improvement programming decision processes is that many people get involved. As we mentioned earlier, lots of people are interested in setting priorities for budgets. However, group-based GIS is a relatively new area of research, and it is in this area that progress with GIS-based improvement programming applications can be made. We mention a case study in this chapter dealing with transportation. Transportation improvement programming is probably the "stalwart" application of all capital/resource improvement programming, because there is so much money involved in transportation improvement across all communities. This chapter reflects that emphasis, but we go beyond the transportation domain and also discuss improvement programming in the context of land and water resource domains.

In the material to follow, we first provide an overview of improvement programming in Section 10.1. In section 10.2, we provide a comparative view of three organizations that undertake transportation improvement programming, then synthesize those approaches relative to decision workflow. In section 10.3 we turn to three case studies—one each in land, transportation, and water resources—to provide insights about how GIS can be useful in improvement programming.

10.1 Overview of the Workflow for Improvement Programming GIS-Based Analysis

Improvement programming analysis is embedded within an overall process of improvement programming. For example, the New Jersey Statewide Transportation Improvement Program (STIP) serves two purposes (New Jersey Department of Transportation 2006). First, it presents a comprehensive and self-contained guide to major transportation improvements planned in the state of New Jersey, therefore providing a valuable reference for implementing agencies (e.g., the New Jersey Department of Transportation and the New Jersey Transit Corporation) and all those interested in transportation issues in this state. Second, it serves as the reference document required under federal regulations for use by the Federal Highway Administration and the Federal Transit Administration in approving the expenditure of federal funds for transportation projects in New Jersey.

Federal legislation requires that each state develop one multimodal STIP for all areas of the state. In New Jersey, the STIP comprises a listing of statewide line items and programs, as well as the regional Transportation Improvement Program (TIP) projects, all of which were developed by the three metropolitan planning organizations (MPOs). The TIPs contain local and state highway projects, statewide line items and programs, as well as public transit projects.

New Jersey is completely covered by three MPOs: the Delaware Valley Regional Planning Commission, the South Jersey Transportation Planning Organization, and the North Jersey Transportation Planning Authority, Inc. The New Jersey STIP includes the three MPO TIPs without modification. Aggregating the MPO TIPs is a matter of convenience to allocate federal funding. The U.S. Department of Transportation (DOT) deals with allocating funds to 50 states rather than to the hundreds of MPOs spread across the states, leaving the states to pass the money on to MPOs that conform to federal regulations.

The New Jersey STIP conforms to—and in many cases exceeds, as its website contends—the specific requirements of the federal regulations, including the following:

1. It lists the priority projects programmed for the first 3 years of the planning period.
2. It is fiscally constrained for the first 3 years. A detailed discussion is often provided within the TIP.
3. It contains all regionally significant projects, regardless of funding source.
4. It contains all projects programmed for federal funds.
5. It contains state-funded projects.
6. It contains expanded descriptive information—considerably more than is required by the federal regulations.

GIS-based data analysis for improvement programming focuses on prioritizing among different projects, some of which then get included in a TIP, and others of which do not, due to lack of funds. Not all impacts are well understood, as in project implementation (supposedly and eventually). The lack of impact information is due to the lack of available time to gather data about projects: There are simply too many projects and too few resources most of the time. Consequently, the basic challenge is to select among the projects to be budgeted given what people know or what they at least are willing to pay to know.

Until 2002, King County, Washington, presented a number of indicators associated with improving the quality of life in the county. The county used at least 45 indicators across a number of issue areas that had once achieved some level of concern. However, in 2003, King County policy staff moved to a more conservative perspective, showing only indicators that could be measured continually—because indicators are only good when compared over time. Some of the policy indicators were about land use change, which in an urban context means that "improvement programming" is at work.

The basic process for improvement programming is as follows:

- Establish a funding mechanism.
- Prepare criteria (to include funding constraints).
- Enumerate alternatives based on relevant criteria.
- Perform an evaluation.

Perhaps the biggest challenge related to the improvement programming decision situation is for an organization (community) to raise sufficient funds so that many of the projects in the mix can be funded. Budgeting is a process that all organizations undertake every year to establish "work programs." Capital improvement programs are the work programs of the community. They are the foundation of quality-of-life concerns in communities.

10.2 Comparative Perspectives in Improvement Programming Decision Situations

A wide-variety of land use development programs exist—so many that it is fair to say the differences are more plentiful than the similarities among organizations and jurisdictions. Much of this difference has to do with what people, organizations, and jurisdictions do with land. Nonetheless, the commonality among them all is that there is never enough money to satisfy all needs; thus, priorities must be set. A sample case is provided in section 10.3.1 on housing.

Transportation improvement programming is not only one of the most visible activities in regard to the results of improvement in an urban context, but also invisible in terms of decision situations. It is visible in the sense that more people know about the process within organizations because of its impact on society. It is invisible because few people in the general public know how it occurs.

Everyone is affected by transportation change. Improvement programming occurs on a number of scales. A study examined three scales of transportation improvement to compare decision processes to uncover similarities and differences (Nyerges, Montejano, Oshiro, and Dadswell 1998). Each scale was associated with a different organization in the central Puget Sound region: Duwamish Coalition, Puget Sound Regional Council, and King County DOT. Hence, there were at least three different decision situations, all within the improvement programming category.

From the results of an early decision situation assessment of the decision processes among the organizations, a task model based on an early version of enhanced adaptive structuration theory (EAST; see Figure 10.1) was created. Boxes 1–3 depict the convening influences on the process. Software technology is depicted in Box 1. The institutional rules for setting the stage for the process (i.e., laws) are depicted in Box 2. Group interaction rules (social courtesy) are part of the issues in Box 3. The first and third boxes were flipped in the transition from EAST1 to EAST2. The reason for the change was that in the original version of adaptive structuration theory (AST), DeSanctis and Poole (1994) used advanced information technology as the introduction to *structuration*, because information technology was the focus of study. However, Jankowski and Nyerges (2001a) revised EAST, introducing the idea that laws are the main structuring aspects of information technology use. EAST was revised accordingly. Most people, as well as organizations, worry about being on the "wrong side" of a law. Organizations and people use the breach of law as reasons to *sue* each other. Box 4 describes the

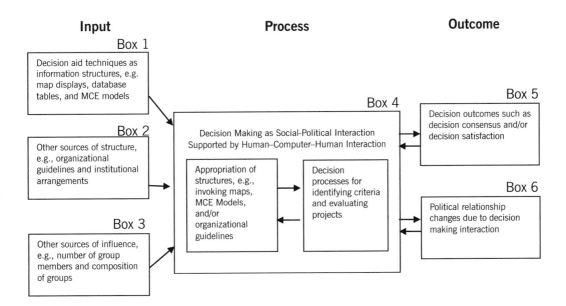

FIGURE 10.1. A task model for summarizing the influence of group-based GIS technology in support of TIP project selection. From Nyerges, Montejano, Oshiro, and Dadswell (1998). Copyright 1998 by Elsevier. Reprinted by permission from Elsevier.

interaction process. People appropriate technical information in the form of tables and models, then manage the decision processes as a series of tasks. Boxes 5 and 6 describe the outcomes of the process. Decision outcomes are perhaps most visible (i.e., they represent the package of projects to be included in an improvement program). However, social outcomes are also part of the process. People develop social relationships, which can take the form of groups of interest or coalitions that help to see projects through to their implementation.

That task model leads to the synthesis of a sequence of tasks described in terms of eight phases (see Table 10.1). We concluded that the improvement programming decision situations are more similar than they are different (Nyerges et al. 1998), but there are some individual nuances. The differences are not substantial enough to make a difference to GIS-based decision support.

We enumerated the software capabilities needed for each task phase using a systems capabilities framework (see Table 10.2). There are many capabilities that "commercial, off-the-shelf software" does not implement, because many organizations are not yet demanding group-based capabilities in these versions of the software. However, many organizations are customizing GIS software to support their decision processes in specific ways. We introduce a decision experiment as the case study in section 10.3.2. project alternatives)

10.3 Improvement Programming Case Studies

In this section we examine three case studies for improvement programming: affordable housing development, regional freight transportation mobility, and water supply and drainage improvement.

TABLE 10.1. Phases in a Decision Process for TIP Improvement Site Selection

Initial screening of projects

- Phase 1: Establish basic criteria for preliminary project screening based on mandates.
- Phase 2: Identify a comprehensive list of projects collected from various sources as suggestions for capital improvement.
- Phase 3: Screen projects according to the basic criteria established in Phase 1.

Refined evaluation

- Phase 4: Refine criteria to evaluate the projects passing the initial screening process.
- Phase 5: Describe aspects of the projects in more detail, using the criteria and weights from Phase 4, as well as any other concerns that need to be made known.

Selection of projects

- Phase 6: Consider unusual circumstances in the evaluation of the projects.
- Phase 7: Compare the project alternatives to each other to determine those most worthy of consideration.
- Phase 8: Negotiate for the most important projects.

TABLE 10.2. Decision-Aiding Techniques for a Group-Based GIS for Transportation

Level 1: Basic information-handling support

a. Information management: storage, retrieval and organization of transportation data and information (e.g., distributed database management system support)

b. Representation aids: manipulation (analysis) and expression (visualization) techniques for a specific part of a transportation problem (e.g., shared displays of charts, tables, maps, diagrams, matrix and/or other representational formats)

c. Group collaboration support: techniques for idea generation, collection, and compilation; includes anonymous input of ideas, pooling and display of textual ideas, and search facilities to identify possible common ideas (e.g., data and voice transmission, electronic voting, electronic whiteboards, computer conferencing, and large-screen displays)

Level 2: Decision analysis support

d. Process models: computational models that predict the behavior of real-world processes (e.g., simulation models of traffic congestion, or air pollution contribution based on traffic volume)

e. Choice models: integration of individual criteria across aspects or alternative choices (e.g., multicriteria decision models using multiattributes and multialternatives for systematically weighted rankings or preferences)

f. Structured group process techniques: methods for facilitating and structuring TIP decision making, (e.g., automated Delphi, nominal group techniques, electronic brainstorming, and technology of participation)

Level 3: Support for group process consistency

g. Judgment refinement/amplification techniques: quantification of heuristic judgment processes (e.g., Bayesian analysis, social judgment analysis for tracking each member's judgments for feedback to the individual or group, and sensitivity/trade-off analysis for comparing transportation project alternatives)

h. Analytical reasoning methods: perform problem-specific reasoning based on a representation of the transportation decision problem (e.g., using mathematical programming or expert systems guided by automatic mediation, parliamentary procedure, or Robert's Rules of Order; identifying patterns in reasoning process).

Note. Based on Nyerges, Montejano, Ohiro, and Dadswell (1998; adapted from material in DeSanctis and Gallupe 1987).

10.3.1 *Affordable Housing Development as Land Use Improvement Programming*

In a land (use) improvement program, people identify and allocate funds to undertake a number of land use development projects. Land improvement programming often goes by the name of land (use) development, because most people view developing land in urban–regional settings as an "improvement." *Land use development* can mean anything from restoring a degraded habitat (upgrade in currently degraded) to upgrading a building for recreational use or developing new commercial/industrial buildings on land that was once a *brownfield* (i.e., former contaminated land that was cleared of hazardous materials.) "Improvement" is a matter of interpreting one's goals and objectives (plus measured criteria) for an area. Thus, "programming" an improvement means identifying the funds for an improvement among many improvement options, setting priorities in the medium term (i.e., a budget cycle) in terms of the goals, and establishing objectives—perhaps in terms of the long-range plan.

Land use programming issues touch on social, economic, and environmental conditions across the community. Many urban problems stem from social, economic, and environmental processes intermingling and thereby creating external impacts among the conditions. Nonetheless, given the decision autonomy that people seek within a democratic society, we move forward with the best of intentions based on the goals and objectives of our separate perspectives. As a result, there are many different perspectives on growth conditions/problems. A reader might ask whether any of the following situations are happening locally in communities: Is rapid growth within "bedroom communities" generating traffic congestion and fiscal shortages, as housing outpaces employment? Do communities with high-perceived livability have high housing prices that block moderate- to middle-income families (i.e., little availability of affordable housing)? Are older suburban communities experiencing declines in business and neighborhood conditions; while firms and residences move to new locations, perhaps in the suburbs? Are residents of central cities facing reduced job opportunities, increased crime, and deteriorating and/or poor housing conditions? Are rural areas losing jobs and people as residents shift to more urban areas? Do regional-level revenue and tax disparities between rich and poor jurisdictions continue to widen? All of these questions point to events that have been shaping America for decades—worse in some places than others, of course.

Despite the best plans, these conditions/problems can arise from cross-cutting economic and social forces that affect the nation but have their impact within communities. For example, any of the following situations could apply. National and regional forces encourage expansion in certain local areas more than others. Employment shifts from region to region, and encourages population shifts. Employment and population shifts set up challenges for growth management strategies. Over past 50 years, population and employment shifts have resulted in major disparities in economic and social conditions. There are disparities from region to region, community to community, and in jobs and housing. In addition, disparities in communities affect fiscal responsibility to maintain "livability." Furthermore, whereas older areas are at risk of not being able to maintain services, suburban edge areas are in strong competition, new areas face growth challenges. All of these circumstances can challenge land resource planning, but to take action, one must focus on land resource programming.

Land resource programming is a component of growth management programming. There are certain principles to follow in growth management–oriented land resource programming. It is programming that is meant to influence development in several ways. First, it guides economic and social forces by balancing the spread of new development, with efforts to stabilize or revive older neighborhoods. It attempts to modify tax and infrastructure investment policies to influence (local) locational decisions. Second, such programming attempts to mitigate adverse outcomes by improving social conditions through a focus on economic opportunities, job access, housing conditions, investment in public facilities, and services. Third, such programming tries to avoid further inequities by engaging in conscientious public facility siting and investment decisions.

Management of growth and change must link physical development with human development. Land use improvement programming can take the form of (1) affordable housing programs for low- and moderate-income families, (2) economic development program to retain and attract jobs, and (3) redevelopment, infill, and renovation programs to maintain existing neighborhoods and employment centers. We consider affordable housing to be one of the social goals in growth management, but it is also very much a part of land use development and an economic issue.

It is true that zoning for housing has always been a standard approach to growth management; but we all need to be aware that, in desirable places, it actually helps to push housing costs high. Affordable housing programs have been around since the U.S. Housing Act of 1937. Federal monies flowed to local areas until the mid-1970s; since then, those funds have been drastically cut. State, local, and public–private housing programs have picked up the momentum that once was a federal responsibility. Local public–private partnerships are very much alive and use a variety of funding sources, mostly low-income housing tax credits.

Three types of affordable housing programs are common in a growth management context, any of which could be emphasized by the local government. First, there are inclusionary housing programs in which developers are required to incorporate affordable housing units into proposed residential development. Second, there are linkage programs in which developers of nonresidential buildings are required to include some component of residential units in development; this requirement promotes the urban village concept. Third, there are attempts at streamlining regulations to clarify language of ordinances, coordinate procedures, and remove unnecessary duplication.

Washington State has a successful approach to affordable housing, the Washington State Housing Finance Commission (WSHFC; 2006a). The commission is a quasi-governmental organization that fosters affordable housing. It is a publicly accountable, a self-supporting team, dedicated to increasing housing access and affordability. The commission expands the availability of quality community services all across Washington State by focusing on certain location-based criteria that encourage investment. Thus, the commission fosters economic development at minimal costs to the citizens of Washington State. It is a public–private entity that works through the private sector to encourage investment in housing by awarding tax breaks.

We can consider the nature of the program in terms of its potential for GIS workflow by considering the particular activities of the commission with reference to WSHFC website documentation (2006a). Those opportunities are described on the WSHFC website in considerable detail. It is quite amazing how stable the website material has been over the 2001–2006 period during which we have been using that material in the classroom.

The WSHFC (2006b) supports a low-income, affordable housing tax credit program that can change from year to year but generally provides various opportunities across the state as needed. The Low-Income, Affordable Housing Tax Credit Program Application contains a number of references to spatial data relationships and is a very good candidate for geospatial processing (consider the criteria in Table 10.3).

As you consider the criteria in Table 10.3, imagine working for an affordable housing developer and undertaking a GIS project workflow similar to that for the Green County GIS data analysis project outlined in Chapters 3 and 6. It is basically a site selection problem with incredible social payoff.

In the WSHFC program (2006c), some geographic areas are favored over others. The application identifies qualified census tracts in Washington State (see WSHFC website exhibit J; 2006d), but they change from year to year, depending on the local economies across the state. Qualified census tracts are favored for housing development investments in the form of tax incentives (reduction in business taxes) over census tracts that are not on the qualified list. The Low-Income, Affordable Housing Tax Credit Program Qualified Allocation Plan contains several selection criteria for various applications in the particular locations (see WSHFC website exhibit B; 2006e).

The Commission can decline to consider projects that fail to meet minimum standards based on these criteria. Even if the developer characterizes projects, the Commission may determine the scope or otherwise define a "project" or "projects" for purposes of ranking applications, and reserving and allocating tax credit. Thus, there is considerable power and authority in the commission to direct the development of affordable housing across Washington State.

TABLE 10.3. Selection Criteria for Determining Project Priority Support from WSHFC

1. Projects should be located in areas of special need, as demonstrated by location, population, income levels, availability of affordable housing, and public housing waiting lists; WSHFC approved by the Commission on January 25, 2001, Low-Income Housing Tax Credit Program *Qualified Allocation Plan* 2.

2. Projects should set aside units for special needs populations, such as large households, the elderly, the homeless, and/or the disabled.

3. Projects should preserve federally assisted projects as low-income housing units.

4. Projects should rehabilitate buildings for residential use.

5. Projects should include the use of existing housing as part of a community revitalization plan.

6. Projects should be smaller rather than larger.

7. Projects have received written authorization to proceed as a U.S. Department of Agriculture, Rural Housing Service multifamily, new construction project approved by the Commission.

8. Projects are historical properties.

9. Projects should be located in targeted areas.

10. Projects should leverage public resources.

11. Projects should maximize the use of credits.

12. Projects should demonstrate a readiness to proceed.

13. Projects should serve tenant populations of individuals with children.

14. Projects are intended for eventual tenant ownership.

10.3.2 *Transportation Improvement Programming*

As we mentioned in earlier chapters, in the United States, a regional transportation improvement program (TIP) is the formal name given to a transportation program document that contains the transportation project list developed through a regional participatory decision process. A TIP is a 3-year program of projects that must be created (or updated, as the case may be) every 2 years. In the 1990s, regional TIPs were rather different than TIPs in the generations preceding them. The reason is that the Intermodal Surface Transportation and Efficiency Act (ISTEA; 1991) passed by the U.S. Congress, was a major change in federal transportation law from the 40-year history of the National Defense Highway laws that mandated the U.S. Interstate Highway System. The federal law named Transportation Equity Act for the 21st Century (TEA-21 1998) replaced ISTEA in the 1998–2005 period, and promoted most of the same cooperative policies, but extended and refined them further to recognize that transportation is one among many factors to enhance livability. The current law, named Safe, Accountable, Flexible, Efficient Transportation Equity Act: A Legacy for Users (SAFTEA-LU 2005) (2004–2010), continues to emphasize mobility, but safety and equity in transportation systems are of growing importance. All of these laws passed by Congress mandate that every metropolitan area in the United States. organize, in cooperation with state transportation organizations, an MPO to coordinate plans, programs, and projects within a region. Such coordination is required so that an MPO can receive federal transportation funds for transportation improvement, as passed through the particular state department of transportation overseeing the MPO activity. The 1990 Growth Management Act (GMA; as well as enhancements in 1991), adopted by the Washington State Legislature, mandate that as certain contiguous counties grow in population, they must organize, in cooperation with the Washington State DOT, a Regional Transportation Planning Organization (RTPO) to coordinate plans, programs, and projects across the counties to receive state transportation funds for transportation improvement. To avoid duplication, the Puget Sound Regional Council (PSRC) is both the MPO and the RTPO for the central Puget Sound region of Washington State, with primary responsibility to coordinate transportation improvements across King, Kitsap, Pierce, and Snohomish Counties (see Figure 10.2).

The Freight Action in Seattle–Tacoma (FAST) Corridor Program in the central Puget Sound region is an example of a specialized TIP. Projects from the FAST Corridor Program have been added to the regional TIP program from time to time over the past decade, and, as of 2006, nine of the 25 projects have been completed (PSRC; 2006c). Throughout the program, from the beginning in the late 1990s to the present, there have never been enough funds available to implement all projects in any given year. In the early days of the program in the late 1990s, a student group at the University of Washington Global, Trade, Transportation and Logistics Program undertook a study to prioritize projects for enhancement of the overall corridor (Nyerges et al. 1998). As part of the study, the student participants performed a decision experiment using the

FIGURE 10.2. Puget Sound Regional Council four-county region in Washington State. Adapted from Jankowski and Nyerges (2001a). Copyright 2001 by Taylor & Francis Ltd. Adapted by permission.

group-based decision support software GeoChoicePerspectives (GCP) that we developed (Jankowski and Nyerges 2001b). GCP comprises three modules: GeoVisual for individual and group map displays; ChoiceExplorer for multicriteria decision support capabilities; and ChoicePerspectives for accumulating individual vote rankings. The GeoVisual module runs as an extension of ArcView 3.x, whereas the other two modules run as stand-alone applications. Together the modules in GCP support stakeholder negotiations of project selection for any point-, line-, or area-based project base file. To begin the student decision process, a base map with hot links to project photos helped orient student participants rather quickly to various rail- and highway-grade separation locations (see Plate 10.1).

The student group performed stakeholder interviews to identify values and objectives to support their realistic role play. The objectives qualified a set of criteria that were collected by the actual stakeholder groups. Therefore, using several criteria to describe each of the projects, the student group was able use the ChoicePerspectives module to create a multiparticipant ranking of the rail- and highway-grade separation projects that might best address freight movement near and through the FAST Corridor, then display the combined ranks in GeoVisual as a consensus map (see Plate 10.2). The sizes of the circles represent the rank priority of the project: The larger the circle, the higher the rank. The green, yellow, and red colors indicate high consensus (green), medium consensus (yellow), and low consensus (red) among the aggregated, ranked votes.

Maps are about spatial relationships (i.e., the interactions among locations). GIS map displays and query processing capabilities help human beings process large volumes of information in a small amount of time, because the map displays highlight the results of queries. Without the map display driven by query processing, the analyses would add a considerable cognitive load on a software user, perhaps hindering spatial problem solving for some users, and perhaps making a task next to impossible for others. Having tables and map displays connected to multicriteria capabilities can reduce cognitive load substantially, which is why software like GCP is being developed and tested by stakeholder groups.

One way to reduce cognitive load substantially when looking at comparisons among alternative sites is to perform sensitivity analysis (SA). SA can check the robustness of the scoring and weighting. The student group used a multicriteria SA tool in the FAST Corridor project to create displays, such as the table in Figure 10.3 and the histogram in Plate 10.3. The weights of the criteria are depicted on the left side of the Figure 10.3. The effects of small adjustments in weights can be easily detected with the SA display in Plate 10.3.

TIP decision making is an institutionalized process in many communities around the world. In the United States, metropolitan TIP decision situations arise every 2 years, often affecting hundreds of thousands, and sometimes millions of people. However, many other opportunities to use GIS-based decision support tools also occur. Decision experiments like the one we just described are undertaken to raise awareness of how use of geospatial information technologies can promote efficient, effective, and equitable decision support, but it is up to organizations to adopt them for use.

10.3.3 Water Resources Improvement Programming

As we mentioned in Chapter 9, the City of Seattle Public Utilities Department (SPU) addresses water resource planning, programming, and projects. As in other city depart-

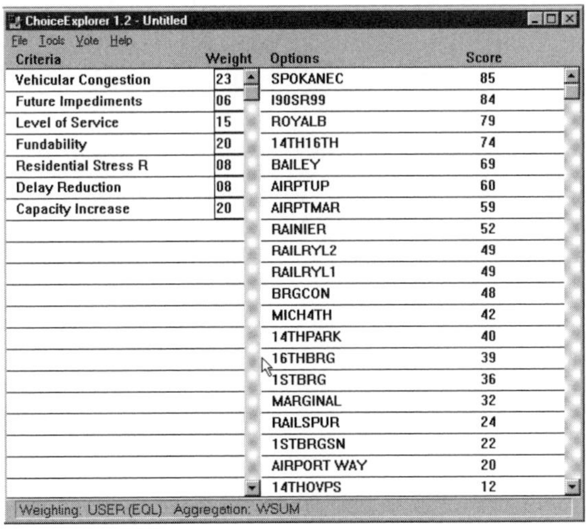

Criteria	Weight	Options	Score
Vehicular Congestion	23	SPOKANEC	85
Future Impediments	06	I90SR99	84
Level of Service	15	ROYALB	79
Fundability	20	14TH16TH	74
Residential Stress R	08	BAILEY	69
Delay Reduction	08	AIRPTUP	60
Capacity Increase	20	AIRPTMAR	59
		RAINIER	52
		RAILRYL2	49
		RAILRYL1	49
		BRGCON	48
		MICH4TH	42
		14THPARK	40
		16THBRG	39
		1STBRG	36
		MARGINAL	32
		RAILSPUR	24
		1STBRGSN	22
		AIRPORT WAY	20
		14THOVPS	12

FIGURE 10.3. Decision table of project ranking tied to histogram of sensitivity in Plate 10.3. From Jankowski and Nyerges (2001a). Copyright 2001 by Taylor & Francis Ltd. Reprinted by permission.

ments, there are a number of subprograms in SPU. Each subprogram contributes capital improvement project suggestions for the capital improvement program.

We look at two capital improvement program contexts in this section. First we consider the water supply plan, and the improvement program as part of that, just to get an overview of how it is associated with water supply (i.e., where the water comes from). Then we turn to drainage (i.e., where the water goes).

10.3.3.1 Water Supply Improvement

Here we have a look at relationships between the short-term Capital Improvement Program (CIP) and the long-term Capital Facilities Plan (CFP), which both comprise a Water System Plan (WSP). The SPU website provides a section by section presentation of the WSP (City of Seattle 2006b). The CIP is part of that plan. The following material borrows heavily from section 8 of the WSP to provide a clear sense of the opportunity for using GIS in water-related capital improvement programming.

SPU prepares both a short-term (6-year) CIP and a long-term (25-year) CFP to ensure that needed system improvements are planned, budgeted, and implemented. The 6-year CIP describes all of the capital projects planned over the near term, with emphasis on the detailed budget of the first 2 years. It is updated every 2 years as part of the city's budget process, with minor adjustments made midbiennium. The most recent update to the CIP was adopted by the City Council in 2005 (City of Seattle 2006d). The 25-year CFP describes all of the capital projects needed over the longer term and is updated somewhat less frequently than the CIP. The most recent complete CFP update was prepared in 1995.

The WSP was prepared with the versions of the proposed 2001–2006 CIP and 2001–2025 projected CFP that were originally made available on April 15, 2000. Excerpts from the proposed 2001–2006 CIP, including descriptions of major projects, are provided in an appendix to the plan. The projected 2001–2025 CFP is included in another appendix to the plan. For this WSP, only the years 2001–2020 are used from the CFP to align with the planning horizon for this document.

In late 1999 and early 2000, SPU staff took part in a departmentwide (across all subunits) effort to identify new CIP projects, as well as changes and adjustments to previously identified projects. Meetings to facilitate this process were held with groups of SPU staff whose work is related to the following categories of projects:

- Conservation
- Dams
- Distribution system water mains and hydrants
- Habitat restoration, habitat conservation plan implementation, fisheries, and Endangered Species Act compliance
- Information technologies and customer service improvements
- Intermittent supplies
- Metering and service connections

- New supply
- Neighborhood planning and other agency relocations
- Operations facilities
- Pump stations, including new pump stations, as well as pump station rehabilitation
- Reservoirs
- Seismic upgrades
- Tanks and standpipes
- Transmission pipelines, including new pipelines, as well as pipeline rehabilitation
- Water quality
- Watershed facilities

Each group created and prioritized a detailed list of new and previously identified projects. General SPU goals used in identifying and prioritizing new projects were as follows:

- Regulatory compliance and public health protection
- Environmental stewardship
- Customer service
- Infrastructure maintenance needs
- Strategic technology implementation
- Neighborhood benefits
- Meeting growing demand

Whenever possible, various project alternatives were considered, and the apparent, most cost-effective projects (based on need, risk, cost, and benefit) were included on the list. The group lists were then merged into a single list and compared with projected funding availability based on estimated rate increases. With only high-priority projects included in years 2001–2006, the proposed CIP exceeded the funding levels presented to the Executive Office and City Council in previous years, so there was the potential that the proposed CIP would be scaled back during the budget adoption process.

The proposed CIP and projected CFP are organized broadly into five categories: water infrastructure, water quality, water supply and conservation, other agency projects, and technology. Each of these categories is described in more detail below.

Water infrastructure projects include efforts to rehabilitate or to replace system components that have either exceeded their useful lives or are beyond repair, or to make improvements that extend the useful lives of assets. This program category includes repairs and upgrades to water mains, pump stations, tanks and standpipes, dams, operations facilities, and watershed facilities. It also includes metering repairs and upgrades, and service connection work. A total of $644 million in infrastructure projects was included in the projected CFP for 2001–2020.

Some of the key projects and programs in this category (with estimated 20-year expenditures given in millions [M] of 2001 dollars) are as follows:

- Distribution system (customer) meter replacement ($33M)
- Service renewals and retirement program ($61M)
- Installation of new taps ($55M)
- Watermain and feeder replacement and rehabilitation ($86M)
- Blowoff improvements on transmission lines ($21M)
- Cathodic protection of pipelines ($8M)
- Rehabilitation/replacement of transmission lines ($185M)
- Seismic upgrades of pipelines ($42M)
- Cedar watershed bridge replacement and road improvements ($8M)
- Booster pump stations in distribution system ($11M)
- Completion of four additional phases (II, III, IV, and VIB) of the Tolt 2 Pipeline ($30M)

Water quality projects are required to protect public health, and to meet state and federal health regulations. This program category includes design and construction of major water treatment facilities, and the program for covering SPU's nine, in-town open reservoirs. A total of $171 million in water quality projects is included in the projected CFP for 2001–2020. Key projects and programs in this category are as follows (with estimated 2-year expenditures):

- Design and construction of ozonation treatment for the Cedar reservoir source ($101M)
- Replacement of Lincoln and Volunteer Reservoirs with new, belowground, covered structures ($19M)
- Rehabilitation and installation of floating covers at the other seven open reservoirs (Bitter Lake, Lake Forest Park, Beacon, Myrtle, Maple Leaf, Roosevelt, and West Seattle) ($45M)
- Replacement of rehabilitation of the Water Quality Laboratory after 2015 ($5M)

Water supply and conservation projects increase the supply of water, protect existing supplies, or reduce demand through conservation. This program upgrades transmission pipelines and promotes residential and commercial water conservation. The projected CFP includes $98 million in 2001–2020 for water supply and conservation projects. Key projects and programs in this category are as follows (with estimated 20-year expenditures):

- Implementation of the Cedar River watershed Habitat Conservation Plan (HCP) ($53M)
- Conservation programs designed to reduce water demand in Seattle's direct and wholesale service areas by 1% per year ($42M)

For new supply, SPU plans to participate in Tacoma's Second Supply Project. It is expected that Tacoma will own, finance, and operate the project, with SPU having the long-term right to a capacity share in it. SPU's share of the capital costs for the project, estimated at $83 million, is not shown in the CFP or the CIP, because the project will be financed by Tacoma. SPU will not own any physical facilities. SPU would make annual payments from its operations and maintenance budget to Tacoma that would cover SPU's share of capital recovery (principal), debt service, and current year operational costs for the project.

Other agency projects include either reimbursable work performed for other agencies or projects undertaken because they are cost-effective when completed in coordination with projects initiated by other agencies. This program category includes projects such as the utility relocations needed to accommodate other agencies' capital improvements. The most significant single agency requesting these projects in the next 20 years will be Sound Transit. In addition to installing pipes to supply water to Sound Transit stations, rail alignment will require the water system to move existing transmission pipelines and water mains. The projected CFP includes $21 million in other agency project costs for 2001–2020.

Technology projects include information technology projects needed to improve SPU's efficiency, productivity, reliability, and customer service. Because of the rapid pace of technological change, a placeholder amount rather than specific technology projects has been included for the period 2011–2020. A total of $77 million in technology projects was included in the projected CFP for 2001–2020. Key projects and programs in this category are as follows:

- Upgrade of the supervisory control and data acquisition (SCADA) system, as recommended by the 1998 SCADA Strategic Plan ($18M)
- Other information technology improvements in areas as diverse as financial management, automated meter reading, and integration of computers with telephone systems ($58M).

10.3.3.2 Drainage Improvement

Much of the material in this section is drawn from the Drainage Program Plan (City of Seattle 2006c), whose mission is to provide cost-effective drainage systems citywide to safeguard public health and property, while protecting our aquatic resources. The 2004 Comprehensive Drainage Plan (CDP) charted a long-term course for drainage in Seattle, with a specific emphasis on 2005–2010 CIPs. The 2004 CDP expands the role of SPU in stormwater management from a conveyance focus to include other elements associated with drainage management, and has created four distinct programs, each with its own goals and objectives:

- Stormwater Conveyance and Flow Control
- Aquatic Resource Protection–Water Quality

- Aquatic Resource Protection–Habitat
- Public Asset Protection

The *Stormwater Conveyance and Flow Control Program* works to alleviate flooding in Seattle, focusing on health and safety, and protection against property damage. As of 2004, SPU has addressed most major flooding problems associated with the trunk system (large transmission pipes), and is working to solve drainage problems locally using detention or infiltration.

The Stormwater Conveyance and Flow Control program goals comprise the following:

- Managing surface water to protect public health and safety, minimize property damage, and protect the environment
- Protecting the value and function of drainage infrastructure and extending its useful life

The *Aquatic Resource Protection program* at SPU has invested drainage funds in projects that reduce or mitigate stormwater impacts on Seattle's aquatic environments through improved water quality, flood control, and habitat conditions. The Aquatic Resource Protection program includes both Habitat and Water Quality sections.

Water quality is fundamental to protecting aquatic resources and public health, and maintaining recreational resources. Seattle's streams, lakes, and marine waters still have water quality problems associated with an urban environment in which contaminants are carried in runoff from streets and other surfaces. For many areas, we still have limited information on the extent of the problem or its source. Proposed program direction in the 2004 CDP expanded water quality monitoring and source control activities.

Urban creeks and shorelines are the habitats of salmon and other water-dependent wildlife. SPU understands the impacts of urban runoff on these habitats, and has worked to protect and enhance water ecosystems. The 2004 CDP outlined an increased focus on habitat, which includes improving and protecting habitat conditions along creeks and affected shorelines, and fostering awareness and stewardship of natural systems and aquatic habitats through outreach, education, and partnerships.

The aquatic resource protection goals for projects include the following:

- Protect and seek opportunities to improve water, sediment, and physical habitat quality in defined key environments associated with drainage and wastewater systems in Seattle.
- Foster awareness and stewardship of water quality, natural systems, and aquatic habitat.
- Create a dynamic and responsive program that can respond effectively to and implement changes necessitated by new regulations, policies, and scientific information.

The *Public Asset Protection program* addresses landslides that present a risk to public health and safety, as well as to public facilities. City departments developed a prioritized list of landslide projects in 1998 as part of the City's overall public asset protection program. The new data on landslides released in 2006 were used to update the list.

Public Asset Protection program goals comprise the following:

- Protect drainage and wastewater infrastructure from undue risks and liabilities due to landslides. This includes projects and programs to reduce risk to vital drainage and wastewater infrastructure, and to educate the public about the risks of owning property in landslide-prone areas.
- Mitigate the direct effects of drainage and wastewater systems operation on or within landslide-prone areas. This includes protecting other properties from landslides that could be caused by inadequate city infrastructure.

In all four capital improvement subprograms within the CDP, a decision process is used, in which prioritization is based on a number of goals, objectives, and criteria. Criteria as "scored attributes" make operational the goals and objectives described earlier in the following ways:

- *Consistency with overall program objectives*: degree to which the project meets overall program objectives.
- *Implementation/phasing*: ease of implementation, and whether the project can be implemented in phases.
- *Cost-effectiveness*: life-cycle costs.
- *Customer service/community support*: level and extent of local and regional support and/or opposition, and whether the project affects a significant number of customers and/or provides for geographic balance within the city.
- *Consistency with other city programs*: including neighborhood plans, watershed action plans, city comprehensive plans, and community and environmental objectives.
- *Environmental stewardship*: the extent to which the project meets the City's environmental goals and/or specific regulatory requirements.
- *Multipurpose use*: whether the project addresses more than one program and/or supports other, ongoing efforts.
- *Use of fund*: project must have a clear connection to meeting our drainage utility purposes as part of improving our system through regulatory requirements, as mitigation of drainage impacts or costs to be shared.
- *Program-specific criteria* will also be considered.

Clearly, there is much potential for the use of GIS integrated with multicriteria evaluation (MCE) in improvement programming at the City of Seattle. Projects can be scored and rank ordered using MCE, and their relative location, rank, and the degree of agreement–

disagreement about projects' merit can be visualized on GIS-generated maps. In addition, changes in project ranks caused by shifting priorities attached to evaluation criteria can be computed and visualized. However, even for such an advanced GIS group as exists at the City of Seattle, the software capabilities to perform spatial multicriteria analysis of projects comprising water resources improvement program are not yet available. That should not stop our readers from encouraging organizations in their local areas to explore the possibilities and benefits.

10.4 Summary

This chapter has focused on the use of GIS in improvement programming. GIS applications for improvement programming are some of the best kept secrets, perhaps because they are so politically volatile. Almost everyone is interested in setting priorities for budgeting money. The real question is, who has the responsibility and the authority to do so? This is why the application is ripe for public participation GIS. In addition, the application is highly interactive—as are most applications that deal with budgeting, and involve give and take on priorities. Given these considerations, group-based GIS is a natural for this kind of application.

Improvement programming occurs on a number of scales. A comparison of three transportation improvement programs in the Pacific Northwest showed how similar they are among organizations and at different scales. With such similarities, the use of GIS for supporting these processes has tremendous opportunity for growth. GIS-based data analysis for improvement programming focuses on prioritizing among different projects, some of which are then included in a TIP, and others that do not due to lack of funds. Not all impacts will be well understood. Data collection about impacts is still a challenge at the level of improvement programming. Impacts are not commonly understood until sufficient time and funds are dedicated to their investigation during project implementation.

Three case studies—on land, transportation, and water resources—have provided insights into how improvement programs are important to various organizations. Land resource programming is a component of growth management programming. We focused on affordable housing for land resource development. A case study for WSHFC shows that there is considerable investment potential in affordable housing development.

We then described the FAST Corridor Program in the central Puget Sound region as an example of a specialized TIP. Over the past decade, there has only been enough money to develop just over one-third of the projects, so there is always a shortfall of money. Such is the case with most improvement programs in all communities. This is why we need to set priorities. GIS-enabled decision support software can be helpful in working through such priorities.

The third case study was about water supply and drainage facility capital improvement at the City of Seattle Public Utilities Department. SPU prepares both a short-term (6-year) CIP and a long-term (25-year) CFP. The six-year CIP describes all of the capital projects planned over the near term, with emphasis on the detailed budget of the first 2 years. The 25-year CFP describes all of the capital projects needed over the longer term and is updated somewhat less frequently than the CIP. The most recent complete CFP update was prepared in 1995 for an ending date of 2020.

10.5 Review Questions

1. What does establishing priorities among projects have to do with improvement programming?

2. What are topical areas for which a county might develop improvement programs?

3. Why might most organizations have been slow to adopt GIS for improvement program–oriented applications?

4. How do analyses for improvement programming–level processes for land, transportation, and water resources compare and contrast?

5. Which improvement programs described in this chapter have been the most prevalent programs to make use of GIS? Why?

6. Describe the basic phases of an improvement program decision process. Where do GIS capabilities fit into the phases you have described?

7. What constitutes a workflow task model for improvement programming–level analysis for housing, transportation, and water resources? Where might GIS be of use in the workflow?

8. How would the freight action strategy improvement program differ from a conventional transportation improvement program? What kind of connection might there be between them?

9. What is the relationship between the City of Seattle Water Supply Plan and the CIP?

10. What is the difference between the Water Supply Program and the Water Drainage Program? Does each really need a CIP? Why or why not?

CHAPTER 11

GIS Data Analysis for Improvement Project Implementation Decision Support

Once we know how much money we have available, and how the money is budgeted by project, then we consider implementing projects within those guidelines. In this chapter, implementation commonly involves three phases: (1) scoping, (2) designing, and (3) building, although we treat workflow in more detail. Those are the same three phases considered for funding approval in an improvement program, because money is allocated for future work in that way.

11.1 Overview of Project Implementation Analysis Workflow

One type of project implementation–level analysis is an environmental assessment (EA). There are two levels, the preliminary EA and a fuller version called *environmental impact analysis* (EIA), that results in an environmental impact statement (EIS). A general description of EA analysis is provided by Ortolano and Sheperd (1995). They describe a number of the opportunities and challenges for such assessments. We can compare the EA process and the EIS process by diagramming the relationship between the two. The preliminary EA process contains only a few steps, but the fuller EIA process, including the EIS outcome, contains many steps (see Figure 11.1). A preliminary EA is used when project impacts are assumed to be small, expected to be few in number, and are not very significant. A preliminary EA report is needed when the impacts are small and thus less significant; whereas the fuller EIS is needed when the impacts are significant.

The general steps of an EIS/EIA include the following phases in line with Figure 11.1 (Randolph 2004 p. 615):

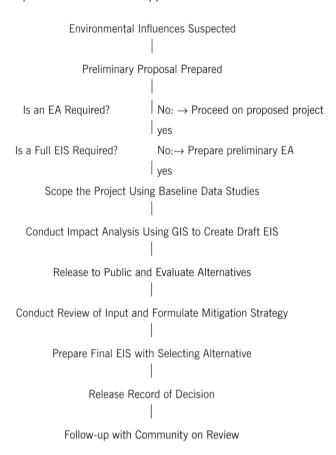

Environmental Influences Suspected
|
Preliminary Proposal Prepared
|
Is an EA Required? | No: → Proceed on proposed project
                          | yes
Is a Full EIS Required?   No:→ Prepare preliminary EA
                          | yes
Scope the Project Using Baseline Data Studies
|
Conduct Impact Analysis Using GIS to Create Draft EIS
|
Release to Public and Evaluate Alternatives
|
Conduct Review of Input and Formulate Mitigation Strategy
|
Prepare Final EIS with Selecting Alternative
|
Release Record of Decision
|
Follow-up with Community on Review

FIGURE 11.1. Relating environmental assessment and environmental impact statement process.

1. *Scoping*: Design the process; draft the work program; identify issues, impact variables, and parties involved and methods to be used.
2. *Baseline data studies*: Collect initial information on baseline conditions and important impact variables, which might include both socioeconomic and environmental parameters.
3. *Identification of impacts*: Concurrent with baseline studies, identify and screen impacts of alternative actions in terms of variables, indicators, and thresholds.
4. *Prediction of impacts*: Estimate the magnitude of change in important impact variables and indicators that would result from each alternative "with and without project analysis." Use project outputs, simple algorithms, simulation models, and/or GIS as needed.
5. *Evaluation of impacts and impact mitigation*: Compare indicator impacts to thresholds; determine relative importance of impacts to help guide decisions; evaluate strategies for mitigation of impacts.
6. *Presentation of impacts*: Present impacts of alternatives in a concise and understandable format.

In King County, Washington, an EIA was performed as part of a multistep analysis process in siting a regional wastewater facility. A four-phase process was used (King County 2004):

- Phase 1: Prepare selection criteria and use to identify preliminary site list.
- Phase 2: Study the selected sites based on conceptual plant layout; six sites need to be identified to move forward.
- Phase 3: Prepare EIS to identify the impacts of the selected sites in Phase 2, and suggest a preferred alternative.
- Phase 4: Conduct permitting and further impact analyses as needed.

The fourth phase is really where project implementation begins. The site has been chosen and the more detailed analysis can be performed for the site.

11.2 Comparative Perspectives on Project Implementation Analysis

As project implementation gets started, a site is assumed already to have been chosen by way of the improvement program, or at least from the multiple sites chosen a single site option will emerge. Sites come in many sizes. A site might be a portion of an acre or several hundred thousand acres. Site size depends on the magnitude of the project—land, transportation, and/or water resource improvement project, or other. GIS data analysis is commonly used when lots of data exist. Such datasets might be developed for very large sites, or the dataset may be large when a small site is to be analyzed in considerable detail.

GIS capabilities for project site investigation have been utilized in the construction industry, particularly in relation to construction management processes. There are GIS applications to support the scoping, designing, and building phases of a project implementation, as well as applications that support an integrated approach. We address each of those phases below.

11.2.1 Scoping Process

After one or more sites have been chosen, the next step is to find out what is known about the site. When scoping a site, a GIS analyst might take advantage of GIS work already performed, for example, in the form of planning studies. Data that might already be available through planning studies perhaps cover a much larger area but nonetheless might be useful.

Site investigation is an important step in scoping a construction project. Existing data are often too coarse when it comes to design. Surface and subsurface conditions influence construction methods and choice of equipment to be used on a project; therefore, they affect the cost and scheduling of projects. Oloufa, Eltahan, and Papacostas

(1994) report on an application of GIS to site investigation. The application makes use of a database for the storage of descriptive soil data, GIS to relate this data to a display of corresponding locations of boreholes, and a graphical user interface (GUI) to facilitate the input, query, and output of data, in addition to drawing bore logs. They show how the two-dimensional (2D) capability of a GIS must move to three dimensions, because there is a need to show soil variation as a function of depth.

Use of existing data at the front end of a project implementation process can save considerable time and resources, as was accomplished, for example, by the Washington State Department of Transportation (WSDOT) in its multiple-project implementation associated with a large corridor in the Seattle metropolitan region. Martinez and Wright (2004) report how GIS can be used to undertake a scoping effort to support "design–build" processes for transportation project construction. The rather large Interstate 405 (I-405) Corridor Project has been under way for quite some time. I-405, the interstate on the east side of Lake Washington in the metropolitan Seattle area, is the second most traveled corridor in Washington State. The corridor supports trips for over 800,000 people each day. It is one of the most significant economic lifelines for Washington State, carrying over twice the weight of goods that are shipped through the Port of Seattle. WSDOT developed an I-405 Master Plan, including roadway and transit improvements, over the past several years. A programmatic Final Environmental Impact Statement (FEIS) for 30 miles of the corridor was complete as of June 2002.

In 2004, project-level designs and environmental assessments were conducted concurrently. More and more infrastructure and facility projects are being programmed as *design–build* projects to expedite the construction schedule and deliver infrastructure solutions more quickly. As such, environmental consultants are finding innovative ways to provide information support for the transportation projects to expedite the construction process using a design–build approach. A *design–build process* is one in which partial design and building occur concurrently. The activity and time associated with constructing major projects has become rather complex and has taken longer in the past decade. Total design, from general design to specific detail design, was becoming overly complex, with considerable design changes occurring. Therefore, the construction industry has moved from total completion of a design followed by total building to a design–build strategy whereby major portions of design, once complete, can direct major portions of building, without getting into fine details that would sometimes change anyway. Complementing project-level designs with project-level environmental assessments assists this design-build strategy.

Two particularly helpful GIS applications were the watershed characterization program and the early environmental investment program. The goal of watershed characterization is "to more completely understand project effects, assess the condition of surrounding natural resources, and identify potential mitigation options that have the greatest opportunity for maximizing environmental benefit while reducing mitigation cost" (Gersib et al. 2004 p. 1). Early environmental investment estimates effects of project construction on the environment, particularly water resource impacts (Martinez and Wright 2004). It is a corridorwide process to identify, rank, select, design, and permit environmental investment opportunities in advance of transportation construction that would create any environmental impacts. Potential early environmental investment (EEI) analysis projects were compiled with the use of basin planning efforts in two water resource inventory areas (Washington State Department of Ecology 2006).

Analysts used three steps to achieve the watershed characterization goal. The first step was to gain understanding of the location and condition of natural resources at both the project site and larger landscape scales. At the project site scale, analysts sought to understand the potential project impacts on existing natural resources. Analysts presented a ranking of existing wetland sites within the project area to assist the project management team in its decision-making process to avoid and minimize impacts to wetland resources.

In the second step, this one at the landscape scale, analysts characterized the condition of key ecological processes (delivery of water, delivery and routing of sediment and large wood, aquatic integrity, and upland habitat connectivity) that the transportation project impacts. Analysts do this by interpreting existing land cover and natural resource data, and by developing databases that identify the location and condition of wetland, riparian, and floodplain resources. Analysts then identify targeted landscape areas with the potential to restore key ecological processes.

In the third step, analysts identified candidate mitigation sites using the wetland, riparian, and floodplain data. In addition to these natural resource datasets, analysts developed a storm water retrofit database to provide additional options for treating storm water in urban areas where few viable natural resource options exist. Analysts established priority criteria, then ranked all candidate mitigation sites for storm water flow control and natural resource mitigation. The storm water flow control priority list was intended specifically to identify potential wetland, riparian, and floodplain restoration sites, as well as storm water retrofit options with the potential to mitigate stormwater flow control impacts of the transportation project. The natural resource mitigation priority list provided a project management team with options for the mitigation of wetland, floodplain, and habitat mitigation needs of a project.

Once a watershed characterization is completed, an EEI analysis can be performed. The EEI uses a watershed approach to identify opportunities for improving aquatic resources, fish habitats, wetlands, water quantity, and floodplains. In order to select potential EEI projects, analysts used GIS to estimate impacts based on the latest road design information.

GIS spatial analyses were performed for right-of-way, noise, and resource impacts. The data were organized to track datasets sourced from state agencies, counties, and local municipalities. As the large datasets were further developed with attributes and clipped, the data were organized by road project and environmental discipline categories. The map documents were also organized by road project and discipline.

Spatial analyses were performed iteratively as more questions arose about impacts on water resources. Working with stream, wetland, fish barrier, drainage, and other resource planning data, GIS helped to guide and quantify design impacts and mitigation areas. The political and funding climate made it necessary for the I-405 team to deal with a multitude of road design and mitigation options. Without GIS tools for spatial analyses, topographical analyses, and data coordination, the program would not have had the flexibility to address options with a quick turnaround time. Overall, the I-405 team comprised five subteams, each addressing one of the five sections of the corridor. Consequently, the map book was quite large. Considerable data sharing was part of the process. Labels and other graphic features were stored separately in geodatabases and used as base mapping for all report graphics.

As another application relevant to the scoping phase of transportation project implementation, Parker, Parker, and Stader (1995) conducted an investigation for the Arkansas Highway and Transportation Department (AHTD) to demonstrate the usefulness of topographic surface modeling, using GIS to perform predictive modeling of soil erosion. The assumption is that terrain, along with other factors, influences the potential for soil erosion in a given area. Disturbance of the land surface in highway construction results in soil erosion and deposition, a source of pollution for streams and lakes. Modeling these various factors with GIS can provide insight for determining the potential for erosion, while construction is in the scoping stage, thereby minimizing the detrimental effects of construction on water quality. The GIS-based application allows for the consideration of erosion prevention strategies such as straw, mulch, or other types of ground cover products designed to prevent or minimize erosion as a result of construction practices. The use of this system allows for effective decisions concerning erosion control before construction has begun and erosion damage has already occurred.

The investigators chose to use the Geographic Resources Analysis Support System (GRASS) for the research. GRASS is a public domain, general purpose, grid-cell-based geographical modeling and analysis software package developed at the U.S. Army Corps of Engineers Research Laboratory (CERL). GRASS databases comprise three major forms: site or point, vector or line, and raster or grid. Although the users of GRASS can model and conduct operations with vector data, GRASS is primarily oriented to raster data. The GRASS system was chosen for this research because it was readily available on the campus of the University of Arkansas through the Center for Advanced Spatial Technologies (CAST) at the time of the study, and because it could easily incorporate predictive models of soil loss as extensions of the spatial analysis capability.

Models to predict soil erosion are available in a variety of forms and can be linked with GIS quite readily. Soil erosion models seek to represent mathematically the actual erosion process. The models tested were the universal soil loss equation (USLE) model, the revised universal soil loss equation (RUSLE) model, Meyer and Wischmeier's simulation of the process of soil erosion by water model, the nonpoint source pollutant loading model (NPS), the watershed erosion and sediment transport model (WEST), and storm water models (SWMs). Each model seeks to predict erosion over a given time interval (i.e., day, month, year). Most of the models incorporate some or all of the factors of the hydrological cycle with some degree of success. The investigators recommended RUSLE as the most beneficial model, because it was most effective at modeling soil loss and, at the same time, easiest to implement. It is readily available from Purdue University (2006).

11.2.2 Designing Process

In an early description of GIS use for construction design, Oloufa et al. (1994) reported how useful GIS databases can be for design–build organizations. They described how organizations can benefit from a single database for building foundation analysis and design, and the resulting design–construction integration. Unfortunately, their application for design is not developed fully; other software developed specifically for those purposes actually do a better job of analysis. Nonetheless, the data management and visualization component of the system was certainly beneficial in the design process.

Combining GIS with simulation has been an effective tool for the construction management process, as shown in other successful case studies (Zhong, Li, Zhu, and Song 2004). A GIS-based visual simulation system (GVSS), composed of simulation and visualization techniques, was developed to improve transparency of complex processes. The GVSS proved to be a useful tool for the design and management of concrete dams. The GVSS offered planning, visualizing, and querying capabilities that facilitate the detection of logic errors in dam construction simulation models. The software also helps us to understand the complex construction process, and it is capable of organizing vast amounts of spatial and nonspatial data involved in simulation. A hydroelectric project on the Yellow River in northwest China was used as an example. An optimum equipment set scheme is determined by simulation of a variety of scenarios taking place under different construction conditions. Likewise, other parameters, such as the construction sequence of dam blocks, the monthly intensity of the concrete process, and the construction appearance at the middle and end of each year, are obtained. Meanwhile, the complex processes of dam construction are demonstrated dynamically using 3D animation, which provides a powerful technique to quickly and comprehensively understand the construction process.

11.2.3 Building Process

Application in support of the building process is among the newer applications of GIS, although some innovative applications have existed for quite a while. The part of construction management that involves building has taken on a systems approach over the past decade. Important aspects of construction management include job activity scheduling and job activity monitoring, among others.

Construction jobs comprise many subcontractor jobs. Job activity scheduling is a critical aspect, making sure that work gets done in a coordinated manner. Although job scheduling is certainly possible (project scheduling software exists) without GIS, people using GIS see more clearly what is being done when a visualization of the process puts it in a spatial context. In March 1997, work began on Seattle's new baseball stadium, Safeco Field. At a cost of more than $485 million, and with an extremely tight construction schedule, this project required efficient and organized execution if it was to be completed on time. To help with Safeco Field's successful completion, Integral GIS, Inc., integrated its construction with the 3D capability of GIS and the 1D time management capability of Primavera Project Planner. This innovation led to the creation of what is now known as 4D GIS Construction Management (Integral GIS, Inc. 2006).

Efficient construction monitoring involves effective communication. In 2006, Newstead, New York, was expanding its water supply infrastructure and sought the help of the engineering consultant firm of Wendel Duchschere (*www.wd-ae.com*), an ESRI Business Partner (Town of Newstead 2006). The most recent, and most extensive, expansion was for Water District 10. District 10 included approximately 60% of the land area of the town. A phased approach was taken for the installation of the water infrastructure because of the size of the water district and the cost of providing water to these residents. Phase 1 included the construction of 28 miles of water line along the most populous roads within the water district. Phase 2, an additional 9 miles of water line, was in the planning stage as of 2006.

The complexities of managing such a large construction project in many different areas of the town led the consultant to develop common construction management practices and look to new technologies, including GIS, to improve the water district's communication and data sharing capabilities. Construction management is fraught with problems when poor and inefficient communication exists among owners, engineers, contractors, and the public. Even in today's technologically advanced society, daily field activities and information recording in cities and towns had until recently been performed manually and delivered weeks or months later. This lack of communication and data sharing has a direct impact on claims, public relations, and project cost containment. To lessen these problems and inefficiencies, the town developed a streamlined and efficient solution for getting the knowledge of the field crew into the hands of the consultant's engineers and town of Newstead staff within the same day.

All documentation is now controlled through a geographically driven interface using GIS software alongside Primavera Expedition, Web-enabled project management software from Primavera Systems, Inc., of Bala Cynwyd, Pennsylvania. The town turned the field inspector's reports and sketches into electronic forms to be filled out and stored on tablet PCs. To ensure a smooth transition from paper to digital form and to minimize training for field inspectors, the town staff re-created the standard hydrant and water service inspection paper forms via database input forms to provide easy, organized access to all collected information.

Each form, sketch, or documented progress photo was input through a customized mapping application combining Newstead basemapping information with computer-aided drawing (CAD) design plans of the water project. Documentation input of all inspected features was initiated by clicking on the desired location in the created map window. Other information, such as daily field reports, material installed by the contractor, and correspondence, is recorded through Expedition. To utilize fully and combine the capabilities of each application, the software was migrated from MapObjects to ArcObjects for a seamless integration with Expedition.

The blend of technology, engineering, management practices, and GIS concepts simplified the transfer and reduced unneeded duplication of information among all parties by organizing the data through linked points on the design plans, and providing wireless data transfer from the field. Each involved party, including the town of Newstead Supervisor and Highway Department Superintendent, Wendel Duchscherer construction managers and engineers, and each construction field inspector, was provided with the same GIS-based interface for tracking construction progress and viewing the documented in-field design and construction issues. That made understanding of construction issues and the construction progress considerably easier.

Reaching out to the public about construction projects is very important in many jurisdictions. Among the GIS applications that have gained considerable attention is one from the Louisiana Department of Transportation and Development (DOTD; 2006). The Louisiana DOTD provides a statewide, online map for people to monitor all proposed transportation and active construction projects in the state by way of a GIS viewer (see Plate 11.1). A user can select among displayable and active layers. There are several tools to query and to explore the data.

11.3 Land Development in Camden County, Georgia: A Case Study

GIS applications for implementing land, transportation, and water resource development and/or construction projects are perhaps among the largest number of opportunities in the growing GIS industry. The application areas are growing as more people find ways to link activity with location. The number of projects in each category is considerable, too large to count. A GIS application about land development is described in the materials below. Designing the most appropriate residential developments on the coasts of the U.S. and around the world will only grow in importance over the coming decades, as more and more of the population moves to the coasts, despite the challenges that climate change might provide relative to a rise in sea level.

11.3.1 Background

The State of Georgia has seen rather large population growth over the past few decades. It makes available a website about growth management, called the Georgia Quality Growth Partnership (2006). The website provides local governments and citizens with the tools and knowledge to transform the way they define, create, and sustain high-quality Georgia communities. As in other places around the country, Georgia communities try to provide for residential growth, foster economic development, and protect natural resources, which requires a delicate balance between the built and nonbuilt environment. A variety of factors come into play, including land values, the abundance of natural resources, real estate market trends, demographics, local ordinances, and community character. As in many high-growth communities, but particularly in coastal communities, information tools such as GIS are needed to help people analyze, visualize, and make decisions about growth and development. That need is the basic premise behind the creation of Alternatives for Coastal Development, a GIS project hosted by the National Oceanic and Atmospheric Administration, Coastal Services Center (NOAA CSC; 2006a). Featuring coastal Georgia, and particularly the area around St. Marys, the GIS project provides information and illustrates results applicable to land development in the coastal zone.

Why is this kind of application generally more important? The coastal zone, comprising a band about 50 miles from oceans and the great lakes, is an area in which over 50% of the people of the United States live. That zone covers just 17% of the land area of the United States, and 13% of the watersheds. Indeed, over one-half the people in the United States live in 13% of the watershed area. Currently, more than half of the U.S. population lives in the 640 counties covering the areas in which major rivers and streams flow into the oceans and the Great Lakes. To put a little more perspective on this, in the 30 years between 1970 and 2000, the number of people living in U.S. coastal watershed counties increased over 34 million—the total population of California. Between 1970 and 2000, population increases within those counties increased their overall population density from 123 to 167 people per square mile. Nearer to the shore, population densities are higher. In areas adjacent to the coast, the population density is over 230 persons per square mile—three times that of the nation as a whole (Colgan 2003). In just the 15 years between 1982 and 1997, 7 million acres of natural land in the same coastal watershed counties were converted to development. Today, almost one-half of the nation's new construction occurs in coastal areas, and over the next 15 years, coastal populations are expected to increase by approximately 27 million people (Pew Commission 2002). The significance of all of this is that GIS can be used to examine the challenges and impacts of most of that development. The need for coastal communities to consider a variety of development scenarios is only going to increase as coastal population densities continue to grow. In this section we look at how a community can consider multiple alternatives for building on a particular site (i.e., what to do with regard to project implementation).

11.3.2 Workflow

The St. Marys GIS project is an example of site-specific development, or what we can call *land development project implementation*. Working in conjunction with the Georgia Coastal Management Program, the Georgia Conservancy, the Georgia Department of Community Affairs, and the City of St. Marys, and with input from additional site and landscape design experts, the staff at the NOAA CSC (2006a) developed and evaluated three hypothetical development alternatives using land, resources, and economic information from coastal Georgia. This project's coastal study site, Camden County, Georgia, experienced a 44.7% population increase from 1990 to 2000, and Georgia is not alone in such growth, as we have described through this book (U.S. Bureau of the Census 2003). Project results are more fully described on the CSC website, and can be used in several ways (NOAA CSC 2006a). First, the website aims to provide specific examples of how alternative development options can impact environmental, economic, and social factors. Second, maps and 3D graphics are intended to help users visualize how alternative design components might look. The project was undertaken to teach coastal communities in general how to put geospatial information technologies to use, and not to assist the St. Marys community with a specific plan. In fact, some of the data came from Charleston, South Carolina, so as not to give local residents the impression that CSC was working for them. As such, it is possible to lay out the following general workflow steps in the project:

1. Conceptual work with state partners
2. Project location
3. Project details
4. Site visitation and data collection
5. Alternative site design scenarios
6. Indicators
7. Review and revise
8. Finalizing scenarios and indicators
9. Creating 3D scenes
10. Final product

• *Step 1: Conceptual work with state partners.* In Georgia, as in many areas, coastal resource managers are grappling with sprawl issues. In developing this project, the CSC partnered with the Georgia Coastal Management Program, the Georgia Conservancy, and the Georgia Department of Community Affairs (NOAA CSC; 2006a). All three organizations provide educational and technical assistance to local communities, and each has articulated a need for GIS-based educational materials to help their constituent groups visualize and make informed choices about developmental alternatives.

The NOAA CSC team worked to define the project conceptually and to select appropriate software to carry out the envisioned scenario comparisons. The software identified by the project was based on a consultation with GIS professionals and on information needs assessment. The project was developed in consultation with experts at the University of Wisconsin Sea Grant and the Land Information and Computer Graphics Facility. Based on a needs assessment for software, the project team selected the main software packages to be ArcView 3.x, ArcGIS, and ERDAS Image GIS and remote-sensing software programs, which are in fairly common use in the coastal management community. In addition, the team used the CommunityViz ArcView 3.x extension for indicator development, the SGWater module of the U.S. EPA's (2006) free Smart Growth Index software for estimating pollutant runoff, and Visual Nature Studio for creating spatially referenced, photorealistic 3D scenes from each scenario.

CommunityViz is GIS-based planning software that allows users to analyze the impacts of alternative site designs in real time, to change assumptions on the fly, and to view quantitative impacts of changes. This project exploited only a portion of the CommunityViz functionality, using it for static indicator development and analyses. Because this project's scenarios were developed via hand drawings outside of CommunityViz, its on-the-fly interactive functionality was not utilized, nor was its 3D visualization component. Rather, in this project, CommunityViz was used for developing and recording project indicators.

The Smart Growth Index (SGI) is a GIS sketch model for simulating alternative land use and transportation scenarios, and evaluating their outcomes using the following indicators of environmental performance:

- Regional growth management plans
- Land use, transportation, and neighborhood plans
- Land development reports
- Environmental impact reports
- Special projects (brownfield redevelopment, annexation, etc.)

Visual Nature Studio (VNS) creates spatially referenced, photorealistic 3D scenes from each scenario. VNS imports GIS data to create a photorealistic image or animation that many people can understand. It provides tools to control visualization directly from GIS data, simplifying and automating the process. A GIS analyst can add as much or as little detail as needed to communicate a design effectively. Imagery can be draped, foliage may be applied, and 3D objects, such as houses and/or street lights, may be added.

- *Step 2: Project location.* CSC relied on the Georgia state-level partner to select a representative study site as a project location. Through the Georgia Coastal Program, a partnership was forged with the City of St. Marys, Georgia, which provided local data and an actual project site currently under development in the area. From the outset of the project, the partners agreed that the project would not portray, endorse, or grade any actual development on the site. Nor would there be any obligation by the town or the developer to implement any project results. Plate 11.2 shows the city of St. Marys and the coast in relation to the study site, and Plate 11.3 shows the study site in relation to the State of Georgia and Camden County.

At this point in the project process, the developer had acquired the site and received the required annexation to extend infrastructure to the property that had no development onsite. Because the project was not involved in early development site identification efforts, no watershed-level analysis of potential development impacts was included or implied in the project. That would have been a planning-level impact analysis in the form of a site suitability analysis.

- *Step 3: Project details.* The team conducted an *audience analysis*, sometimes called a *needs analysis*, to better define project end products and to make sure that partners' needs would be addressed. An audience analysis identifies potential end users and how they might use and benefit from project products. Partners' input to the audience analysis helped to ensure that end products would be tailored to meet the needs of the coastal resource managers and others working to address coastal growth issues. Secondary audiences identified included students, interested citizens, and educators.

The audience analysis took the form of a series of questions for all of the project partners to fill out and discuss. The questions were organized as a step-by-step process to help identify primary and secondary audiences for the project. The website provides an example of the audience analysis used (NOAA CSC 2006b). An audience analysis should always be tailored to suit the needs of a specific project, so analysts might have to redevelop the material to suit their own needs.

A small prototypical dataset was developed with the use of readily available spatial data layers for Charleston, South Carolina, to test the method and software proposed for the project. The prototype helped to ensure that the selected software would be appropriate for the desired project analyses and provided an initial test site as technical staff became familiar with the software (using the prototype to test and document project indicator formulas and required inputs). The prototype helped the team evaluate the feasibility of measuring a range of environmental, economic, and cultural impacts of the development. Appropriate GIS data layers were assembled for the small test area and used to help the team document required data layers and the necessary variable inputs for a variety of indicator calculations.

- *Step 4: Site visitation and data collection.* Center staff and project partners met in St. Marys, Georgia, in 2003, to tour the green site, and to meet town officials and the development project manager. During the site visit, the project team members were able to discuss data collection methods, explore the opportunities and constraints of the site, and forge stronger partnerships through face-to-face communication.

In coordination with project partners and via the state GIS clearinghouse, CSC technical staff assembled base map data layers for the project area, including digital orthophotos and mapped natural resource layers, such as soils, hydrography, national wetlands inventory data, and coastal hammocks. The developer provided site-specific layers, including project boundaries, regulatory lines, jurisdictional wetlands, and elevation contours. In addition, CSC staff created a data layer identifying forest communities on the project site, based on photo interpretation of the 1999 orthophotos.

- *Step 5: Alternative site design scenarios.* In a workshop held in early 2003, hypothetical scenarios were created, based on real development trends in the country. The project team decided to limit the site designs to three scenarios to simplify the analysis and ensure that the quantitative and qualitative results would be meaningful. One project goal was to achieve results showing variation in how different site designs can impact the environment, the economy, and the community. A final decision was made to base the scenarios on conventional, conservation, and new urbanist development. None of these designs represented the actual design under consideration at the project study site, but the scenarios were clearly possible in an application of this kind.

Workshop participants came from diverse backgrounds, and some with land use and planning expertise. Participants broke into three groups to develop the site designs, with at least one experienced land planner in each team. The designs were initially conceptual to make sure that each group included some principles of smart growth, and to ensure that comparison of the three designs would result in meaningful differences among the scenarios. Each group further refined its scenario by drawing lot lines and adding any missing features that would typify the associated development trend. Each group also developed a list of features highlighting key aspects of their design. These features were used in selecting the final suite of indicators to be measured.

Early on, staff wanted the three scenarios to result in the same number of dwelling units, so they could conduct an easy comparison across all scenarios. This became a challenge during site design, so they decided that the number of dwelling units needed to be close but not equal. As a recognized limitation, scenario designs do not always allow equal comparison in terms of how many people each scenario may house.

Hard copy base maps for the project site were provided at the workshop. Each scenario group had a 1:2,400 scale base map and tracing paper overlay sheets for drawing scenario designs by hand. Once the draft scenario designs were completed by each of the subgroups, CSC staff scanned, georeferenced, and digitized each scenario for use in a GIS.

The scenario drawings were scanned with an Ideal Scanner at approximately 250 dots per inch (DPI), and the images were saved in a TIFF format. Using the Spatial Adjustment tool in the ArcMap interface (ArcGIS 8.3), the images were then georeferenced. The registration was a true georeference, as opposed to results obtained using the rubber sheeting method.

Heads-up digitizing was then used to vectorize each of the features within the scenarios. The features (parcels, roads, trails, etc.) were digitized into separate ArcInfo coverages, then converted to shapefile formats. Based on each scenario group's notes and comments from the workshop, the appropriate attributes were assigned to each feature for use in indicator development.

Project partners and ancillary reviewers evaluated the site designs and provided suggestion for tweaking, as necessary, to ensure that the designs were realistic after the workshop was complete. Then, hard copy site designs were converted into digital format for use in a GIS.

- *Step 6: Indicators.* The team began with a large list of potential indicators that gradually was culled through the scenario and data development process. Final indicators were chosen for their relevance to all three scenarios, their ability to highlight the quantitative differences among the scenarios, and the availability of required inputs (including level of detail of the scenario designs). As the project progressed, the team continued to discuss and to refine the indicators selected for analyzing project results. If the team determined that an indicator added no value to the project, it was dropped. For example, the team originally wanted to consider the relationship of the proposed scenarios to schools to help measure walkability and safety for children in the scenarios. The scenario teams decided not to include schools in their designs, because the site-specific need for a school in this area was not considered necessary, and because no demographic analysis of planned or existing households in the area was to be conducted. As a result, this indicator was dropped from the analysis. CSC provides a page to view the final indicators that were used (NOAA CSC; 2006c) but also recognizes that many other indicators could be used but were beyond the scope of the project.

Once the alternative designs were in a GIS format, CommunityViz software was used to formulate and analyze indicators. With this software, multiple formulas can be developed to measure the quantitative differences between scenarios. Research into appropriate variables and modification of formulas was necessary to ensure that the formula inputs realistically represented the desired measure. It was also often necessary to add new attribute fields to the scenario shapefiles for the indicator calculations. The resulting indicator formulas, and associated variables and constants, are available in the indicators methods portion of the website (NOAA CSC; 2006d). The formulas and variables developed are based on either the opinion of experts or specific references from published literature.

- *Step 7: Review and revise.* Project partners took advantage of several opportunities to provide feedback and to suggest changes to the scenarios via e-mail, at a meeting, and at the services center. Draft digital versions of each scenario were printed on large format maps and displayed for comparison at the meeting. Preliminary results for some of the project indicators were also presented. Additional details for the 3D scenes, such as setback width, sidewalk and trail designation, canopy cover, were documented for each of the scenarios by the groups that created them. Such site design–level details were considered conceptually and only for the purposes of developing the 3D scenes.

- *Step 8: Finalizing scenarios and indicators.* Once the scenario designs were finalized by each of the subgroups, the projects indicators were calculated, producing results for each scenario (NOAA CSC 2006c). As described previously, the project indicators were evaluated using either the CommunityViz or SGI Water Application. Three scenarios are available at the NOAA CSC (2006e) website.

- *Step 9: Creating 3D scenes.* Once the scenarios were in digital format and final indicators selected, CSC staff used VNS to develop a selection of 3D views for each scenario. These 3D scenes help to illustrate the look and feel intended for each design and to highlight differences among scenarios. The CSC used VNS software to create 3D scenes and fly-throughs of the three scenarios (3D Nature 2006). The 3D images and animations presented on the NOAA CSC website were also generated with VNS. A broad array of GIS data was imported and used to place objects and ecosystems accurately within the landscape. The CommunityViz application also has a spatially referenced 3D modeling component called SiteBuilder 3D, but it was not exploited in the project.

VNS is a 3D visualization package that requires a significant level of technical proficiency and computing capacity. The time required to produce complex scenes was reduced by configuring multiple personal computers as render engines and simultaneously rendering individual frames on each of the machines. The 3D component of CommunityViz SiteBuilder enabled users to build photorealistic, 3D, interactive models of land use proposals. Whereas CommunityViz was used to illustrate the differences in the three scenarios from a bird's eye view, the VNS application was used to create a streetscape view from the ground level.

- *Step 10: Final product.* Project results on the CSC website are intended for use in several ways (NOAA CSC 2006a). First, the site aims to provide an example of how alternative coastal developments can impact environmental, economic, and social factors. Second, maps and 3D graphics are intended to help users visualize how alternative design components might look. The website has been structured to allow the user to walk through the project processes, and it provides access to project inputs (NOAA CSC; 2006d), as well as the results from the calculations.

The 10-step process demonstrated that often several software packages are needed to undertake a complex GIS project. Improvement project implementations are often among the most complex projects because of the details to be considered. Such details are a matter of course when a robust workflow process is used, as in the St. Marys case study.

11.4 Summary

Community development project implementation commonly involves three phases: (1) scoping, (2) designing and (3) building. These phases are the same in an improvement program. When scoping, one type of project implementation–level analysis is an environmental assessment (EA). There are two levels, the preliminary EA and a fuller version that results in an environmental impact statement (EIS). We described EIS analysis and presented an overview of one associated with regional wastewater facility development. We detailed the steps for scoping, designing, and building each phase potentially making use of GIS.

A 10-step process was described for GIS workflow relative to community development projects undertaken by the NOAA CSC. The 10 steps include, in general, the following: working with partners, scoping out the projects, developing technical indicators, and getting "buy-off" about the products created. GIS can be used in all steps, but it might not be the only software needed to address complex decision problems, as demonstrated in the St. Marys case study.

11.5 Review Questions

1. What constitutes a workflow task model for project implementation–level analysis in terms of an EIS assessment, and how does it differ from the simpler process of EA? Why would an analyst perform one in contrast to the other?

2. What are the comparative differences in scoping, designing, and building steps within a project implementation–level analysis?

3. Describe the GIS activity that can be performed as part of the scoping, designing, and building steps associated with project implementation–level decision support.

4. What are the advantages of using 3D displays in comparison to 2D displays for project implementation–level decision support?

5. What is the goal of watershed characterization? How can GIS help with the process?

6. What constitutes a workflow task model for land use project implementation–level analysis (e.g., in Camden County, Georgia)?

7. Why are partners important in GIS projects like those in Camden County, Georgia?

8. Why is growth management using GIS important along coastal areas?

9. Describe the advantage of alternative site design scenarios and the use of GIS to develop such scenarios. Why are scenarios useful?

10. Why might 3D scenes be important for depicting the alternative site design scenarios?

PART IV

Using GIS for Integrated Decision Support

CHAPTER 12

GIS-Based Integrated Analysis across Functional Themes

W_e present three decision situations to make a case for integrated analysis. Rather than try to address the relationship among the three functional themes that are integrated pairwise then multiplied by three decision processes of planning, programming, and project implementation to address nine functional situation combinations, we have chosen to focus on watershed planning, transportation improvement programming, and a land development project associated with wastewater facility siting. Section 12.1 addresses watershed planning as an integration of land use and water resource planning. In section 12.2 transportation improvement programming is addressed in terms of a connection between land use and transportation activities. Section 12.3 addresses wastewater facility siting, dealing with land use, transportation, and water resources connections.

Integrated analysis across functional themes characterizes a primary thrust of data analyses for growth management approaches. Taking into consideration values and criteria in two or more functional categories is a cross-theme analysis and essentially an extension of the basic analyses approaches in Chapters 9, 10, and 11. We maintain the depth of analysis but broaden the analysis to consider impacts other than within a single theme.

12.1 Work Plans for Integrated Watershed Planning Analysis

In this section we first present an overview of integrated watershed planning analysis that links land use and water resources decision situations. We then discuss more details of the data analysis to be undertaken.

12.1.1 *Overview of a Watershed Planning Decision Problem*

Since the early 1990s, integrated water resource management (IWRM) has received increased attention as a modified form of the more traditional approach to planning (Dzurik 2003). The integration is often between themes such as land use and water resources, because land use has considerable impacts on water resource quantity and quality. Integrated planning is not as broad as comprehensive planning, but it is not as narrow as functional planning. Although there are many similarities to the rational planning model, there are also a number of differences.

In regard to river basin planning, IWRM emphasizes an approach whereby:

1. Water resources have various physical aspects (e.g., surface, ground, quantity, quality).
2. Water is a system, but it is also a component that interacts with other systems (e.g., interaction between land and water, interaction between river and estuary).
3. IWRM can consider interrelationships among water and social and economic development (e.g., role of water in hydropower, industrial production, urban growth).
4. IWRM can consider the river in terms of not only the water itself but also the biological resources that rely on it in its natural state (e.g., fish and wildlife, benthic organisms, plants).
5. IWRM can incorporate a river in its full extent, from headwaters to the estuary, and in consideration of the entire range of potential uses over its length.
6. IWRM can view the resource and its uses from both long- and short-term perspectives (Dzurik 2003; Mitchell 1990).

Despite the need for integrated water resource management, there are several barriers to implementation. Dzurik (2003 p. 107) summarizes the 24 barriers to integrated environmental management presented by Cairns (1991), and articulates the 10 most salient barriers:

1. Integrated management takes time, and time means money; agencies do not fund necessary time for this activity.
2. Turf battles run rampant in organizations.
3. Many participants are unwilling to compromise.
4. Changes in lifestyle required by the integrated resource perspective are strongly resisted not only by some individuals but also institutions and corporations.
5. Society is oriented toward growth rather than maintenance.
6. Political process is oriented toward polarizing issues rather than integrated management.
7. Institutions of higher learning do not train people to think in an integrative manner.

8. Short-term profits are enticing.
9. An attitude of "What has posterity done for me?" is common.
10. Participants are reluctant to change ways of doing things.

Chapter 11 of the U.S. Geological Survey (2009) *National Handbook of Recommended Methods for Water Data Acquisition* describes a variety of water use themes that can form the basis of integrated data analysis for water resource management including: domestic residential, commercial business, industrial, mining, irrigation, livestock and animal specialties, thermoelectric power generation, hydroelectric power generation, and wastewater collection and return flow.

12.1.2 Work Plan for Integrated Data Analysis for Watershed Planning

The U.S. Environmental Protection Agency (2009) watershed academy offers a three-stage process for watershed planning process (See Figure 12.1). These stages are generic in character. The data analysis process could work for any or all of the water resource categories listed as part of the U.S. Geological Survey (2009) items presented earlier. How the specific categories are defined frames the details of what is actually done in an analysis as part of the plan development stage.

Much of the water resource analysis uses an environmental assessment approach to address water use impairments. There are many ways to conduct an environmental assessment; at least a dozen ways have been identified (Vanclay and Bronstein 1995), including environmental impact assessment, social impact assessment, technology assessment, policy assessment, economic assessment, health impact assessment, environmental auditing, and sustainability assessment. According to Heathcote (1998 p. 187), a simple assessment to identify impacts related to water use impairments involves the following steps:

- Focus on a small number of impaired uses.
- Use a small number of indicators to evaluate improvements in impaired water use.
- Make a comprehensive inventory of sources.
- Identify key sources of data.
- Eliminate infeasible options using systematic evaluation techniques.
- Use present and future scenarios to capture likely trends over time.
- Focus on specific outputs, including recommendations for immediate action, deferred action, and additional data collection and analysis.

The simple assessment described here can be performed using Phases 1, 3, 5, 6, and 7 from the synthesized workflow process described in the third column of Table 3.5 in Chapter 3. It is adequate to provide a general characterization of impairments. However, a detailed assessment requires more. A detailed assessment adds Phases 2 and

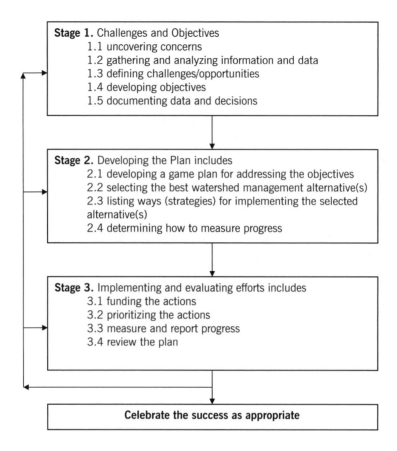

FIGURE 12.1. Three-step watershed planning process. Based on description in U.S. Environmental Protection Agency (2009).

4—process simulation and change related to impacts on alternatives—to the other phases performed in a simple assessment. In a detailed assessment, the analysis will likely yield

- Identification and detailed characterization of specific sources of problem.
- Quantitative evidence in regard to the performance of different management alternatives.
- Elucidation of processes, and resulting cause-and-effect relationships within the basin.
- Detailed and quantitative projections about the impact of specific remedial measures on in-stream hydrology, water quality, and biological systems.

A simple assessment commonly involves mapping with existing secondary source data. The more detailed assessment would involve mapping with data captured in the field.

12.2 Work Plans for Integrated Transportation Improvement Programming Analysis

In this section we first present an overview of an integrated approach to transportation improvement programming analysis that involves the link between land use and transportation. We then turn to more details of the data analysis to be undertaken.

12.2.1 Overview of Land Use and Transportation Concurrency Management Decision Problem

In the land use and transportation decision problem, the goal is to support growth where land use permits will have the least impact on degradation of transportation movement. Because of the diversity of development in the county, a major focus in King County's transportation planning process is to determine the adequacy of transportation facilities in meeting projected auto and transit travel for activity centers and suburban–rural areas of the county. To establish consistency for transportation planning in cooperation with the PSRC, the County makes use of the TAZ coverage provided by the Puget Sound Regional Council (PSRC) as the basis of its transportation modeling. With the use of a GIS, the county splits or aggregates transportation analysis zones (TAZs) where necessary into small-area zones (SAZs) to reflect the character of subareas requiring more refined attention than at the PSRC four-county scale. This is accomplished on the basis of area, network density, and total population, in a manner similar to the creation of TAZs. The SAZs are used by the county's transportation modeling group to forecast travel demand for auto trips. The forecasts are then used to create a transportation adequacy measure (TAM) for each SAZ, as well as for monitored corridors. Consequently, the TAM establishes an operational monitoring link between land use development and transportation improvements.

The TAM is used to assess the amount of congestion in small areas (rather than just on road segments), as well as in corridors, because land use generates trips. Making the connection operational through policy comes in the form of a concurrency management program. King County adopted a concurrency management system, which ensures that adequate transportation facilities are available to meet the requirements of new development in the county (King County 2002a). A concurrency review must be completed by anyone who intends to apply for a land development permit in unincorporated King County.

12.2.2 Work Plan for Land Use and Transportation Concurrency Management

Many of the numbers to implement a TAM are computed with the use of EMME/2 software, with data managed in GIS. The TAM is computed using trips, vehicle miles traveled, and volume-to-capacity ratio. The number of trips on a segment is computed for each zone relative to the total number of trips across the entire county. Computed trips depend on how close or far vehicles are from a zone. The proportion of trips grows smaller as distance from a zone increases.

We use the trips to compute vehicle miles traveled (VMT), as follows (King County, personal communication, May 24, 2006):

VMT = segment length multiplied by the number of trips on a segment

Capacity for a highway is based on the number of lanes, speed, and composition (although speed and lanes are usually sufficient). The volume is the actual number of vehicles at P.M. peak load time on the segment. Thus, we compute the volume-to-capacity ratio (V/C) as

V/C = volume on a segment divided by capacity on a segment

Thus, the TAM for a SAZ is computed as a relationship between VMT and V/C:

TAM = sum (VMT * V/C) across all links divided by the sum of VMT across all links.

A threshold TAM is used as a policy constraint, such that SAZs and monitored corridors should not exceed a certain amount of traffic congestion. Exceeding the threshold amount means that there is congestion (lower mobility) in the area because of the level of land use development or the absence of transportation improvements. The threshold TAM is based on a level of service (LOS) recommended by the Federal Highway Administration for urban highways. LOS is commonly stated as a letter (ranked A through F): A is best and stands for freely flowing traffic; F is worst, as in almost standing still—fully capacitated (Transportation Research Board 1985). LOS is essentially a ranked flow V/C ratio. Consequently, in the case of urban traffic, King County specifies E as 0.99, D as 0.89, and C as 0.79 V/C, whereas for rural V/C, B is always specified as 0.69. Thus, the TAM is to be interpreted as a performance measure; it is an indication of people's mobility on the transportation system. Areas expected to be mobility deficient can be computed based on changes in land use that generate trips—the concurrency link described below. It is important to note that the TAM is an area-based transportation summary that is computed for SAZs, as well as for corridors. It is translated to the same LOS coding scheme that is commonly used in assessing links, but the county feels that link-based congestion interpretations miss a broader level of planning whereby adjacent land uses in small areas feed link congestion (Transportation Research Board 1985).

The TAM (and associated LOS), compared to a threshold TAM, is the operational concurrency link between land use development and transportation development; it guides the approval process for land development permits as part of growth management planning. Applicants for development permits must obtain a certificate of transportation concurrency (or availability) prior to obtaining a development permit. The certificate confirms and establishes the availability of transportation facilities (or supply) to serve the development and commits the capacity to the development. A certificate is not issued if the development causes a violation of transportation LOS, and if no financial commitment is in place to complete the improvements within 6 years.

The King County Concurrency Management Program updates a GIS map of TAM levels for SAZs and corridors twice a year (see Figure 12.2). The light gray shade indicates that the computed TAM is less than the threshold TAM; thus, developing land is not likely to affect transportation flow very much. A medium gray color indicates that a precautionary level of development has been reached. A dark gray color indicates that land development permits will not likely be issued, because the roads are already at, or over, capacity. Areas in white are incorporated cities that manage land use and transportation concurrency according to their own strategy.

There is actually a two-stage test for concurrency—the SAZ check, then the corridor check. If a development permit is denied after checking the TAM in the first phase (i.e., against the TAM for a SAZ), then the permit is denied overall. If however, the permit passes the first-phase TAM check, then the second-phase test is administered (i.e., to check the TAM in the corridor). Assuming the test passes this stage, the permit then moves through the regular permitting process for approval–denial consideration.

A GIS is used to characterize corridors and transportation service areas (based on SAZs) in terms of road and transit service, whether they are inside or outside the urban

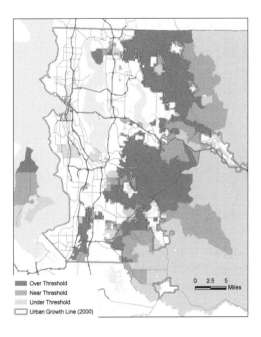

FIGURE 12.2. Concurrency transportation management in King County, Washington, using a transportation adequacy measure based on level of service. Adapted from King County (2002b). Adapted by permission.

growth boundary, and whether they are in an incorporated or an unincorporated area. Together these characteristics help to direct strategy for transportation improvement program projects as part of the capital improvement program.

The King County Department of Transportation (KCDOT) makes use of Community Advisory Boards when creating the Transportation Needs Report (TNR). Furthermore, the advisory boards have provided input to the Transportation Improvement Program projects for their particular areas as well. Interactive GIS has not been used at these meetings, but maps from GIS have been used. The KCDOT is considering use of GIS to enhance public involvement in regard to obtaining public input on the link between the TNR and the Transportation Improvement Program. A strategy to implement that feedback has not yet been established, however.

12.3 Work Plans for GIS-Based Integrated Analysis for Improvement Projects

In this section we first present an overview of an integrated analysis for improvement projects that links land use, transportation, and water resources. We then turn to details of the work plan for data analysis that can be used to implement integrated analysis for improvement projects.

12.3.1 Background on the Decision Problem about Land Use, Transportation, and Water Resources

In this section, we describe an example of a regional land, transportation, and water resources (LTWR) decision problem dealing with siting a wastewater facility. The goal of the integrated LTWR decision problem is to achieve a siting goal and, at the same time, minimal impacts to the surrounding community during the build-out and long-term maintenance of the wastewater facility.

12.3.2 Decision Work Plan

Because of the complexity of the problem, the process moved forward in four phases, each phase having a number of subphases (King County 2001, Figure 3, p. 15).

- *Phase 1: Prepare selection criteria and use them to identify preliminary site list.*
  - Establish minimum site requirements.
  - Identify land areas using GIS/parcel information review, committee nominations process, and an industrial lands search.
  - Define engineering and environmental constraints.
  - Conduct engineering and environmental analysis.
  - Draft site screening and site selection criteria developed for King County Council review and approval.

- Conduct Phase 1 site evaluation using the adopted site screening criteria.
- Identify and conduct preliminary evaluation of potential marine outfall zones.
- Initiate public involvement.
- *Phase 2: Study the selected sites based on conceptual plant layout; six sites identified to move forward.*
  - Develop conceptual plant layouts for each site.
  - Define conveyance corridor concept options (near-surface cut/cover and deep tunnel) for each candidate site.
  - Refine marine outfall zones—examine potential diffuser sites.
  - Assemble candidate sites (general plant layout, conveyance, and marine outfalls) for each proposed candidate site.
  - Conduct onsite investigations and reconnaissance of conveyance corridors.
  - Conduct Phase 2 candidate system evaluation using adopted site selection criteria.
  - Continue public involvement activities.
- *Phase 3: Prepare an environmental impact statement (EIS) to identify the impacts of the selected sites in Phase 2, and suggest a preferred alternative.*
  - Develop project proposal for State Environmental Policy Act (SEPA) and National Environmental Policy Act (NEPA) regulations.
  - Conduct public scoping and receive comments on project proposal.
  - Refine project and alternatives.
  - Conduct a Phase 3 candidate system evaluation (additional feasibility assessment).
  - Develop conceptual mitigation plans.
  - Prepare an EIS on Brightwater candidate system alternatives.
  - Continue public involvement activities.
- *Phase 4: Conduct permitting and further impact analysis as needed.*
  - Permit application and other geotechnical engineering follow-up.

The siting process integrates three concerns: marine outfall (water resources), conveyance corridor (transportation), and the physical site (land use). Combining them requires considerable human resources, because integrated analysis is a complex undertaking. The siting was a multiyear process. Completion of the project (build-out) is slated for 2010.

Phase 3 was the environmental impact analysis, in which multiple themes were considered. However, considerable public contention arose during and after Phase 3 and throughout Phase 4, because the people in the immediate vicinity of the chosen site did not feel that they were heard in regard to negative impacts on their community. Thus, a failure to engage the community sufficiently early on in the project, and perhaps not undertaking sufficient technical analysis at a site level, fostered community contention about the final siting of the facility. Although there was substantial community outreach, a big question is "How much is enough at what point in time?"

There is considerable similarity between the GIS data analysis process for traditional project implementation and an integrated project. However, there is a difference in regard to "who" comes to the table to scope, analyze, and dialogue about projects. It can be said that a difference between the traditional project implementation process and an integrated process is the degree to which multiple stakeholder views are integrated into the *entire* process, which in the case of the Brightwater Facility siting process was not followed. The public involvement process occurred at the end of each phase, rather than throughout the entire phase, as in SEPA and NEPA regulations. *When stakeholder participation is left to the last subphase of each phase, considerable analytical work is performed without the benefit of those who would be impacted most directly, the "publics" in the surrounding community.* At first glance, it appears that performing work in that manner is indeed more efficient, but it is likely to be less effective. When called into question, the analysis might have to be done over, which means workflow process efficiency gains are "out the window."

Integrated watershed resource analysis incorporates public involvement throughout the process. With such involvement, the criteria that show whether success was achieved in the siting process are likely to change. Heathcote (1998), Dzurik (2003), and Randolph (2004) all make a major point about enhanced stakeholder/public involvement. In particular, Randolph addresses land use change and transportation change as an environmental (water, air, noise, etc.) change; thus, multiple perspectives need be considered.

In a democratic society, public decisions should reflect broad social values, and changing policy should equally reflect changing values. Implicit in this is the simple notion of pluralism—decision making by the many, not by the few. At the watershed level, the concept is equally valid: Watershed planning, improvement programming, and project implementation as a process for social change should reflect the values of a majority of the community. In this sense, the watershed planning, programming, and project implementation should not be considered as a plan, a program, and implementation, respectively, but as a process in achieving community change. To accomplish this end, all major viewpoints—publics—must be heard, consensus must be built, and the views of the majority reflected in the change must be considered. Yet for a variety of reasons relating to time and resource demands, unwillingness to relinquish authority, or simple ignorance of opportunities, this does not always happen. Thus, there are several ways to set goals for watershed planning processes as a way to address change. One of the best ways to address goals is to involve a diverse group of stakeholders who bring diverse perspectives to what is valued in a watershed. Each perspective can bring a different sense of "problem awareness" to the planning process.

Many public participation specialists adopt Arnstein's (1969) ladder of participation as a way of framing how citizens can be involved in a planning process. The early (lower) rungs on the ladder involve little participation, or what might be called nonparticipation. The later (middle) rungs of the ladder involve degrees of engagement. The latest (higher) rungs involve degrees of citizen empowerment, whereby citizens participate in the actual decision process.

The ideal form of involvement is based on the perceived problem, the community and its values, and the willingness of decision makers to delegate authority. Various factors are important in designing a public participation program. In essence, these factors relate to the concepts of democratic decision making, including sharing of information, building of trust and credibility, relationships between the public and the decision makers, and conformance with preexisting requirements, such as laws and funding availability. Probably the most critical factors in effective public (citizen) participation are the following:

Before the Process Begins

1. Create a process that engenders mutual respect.
2. Clearly state expectations about:
   - Proposed project scope and key issues.
   - The nature and timing of public involvement.
   - Consultation and communication mechanisms.
   - The level of citizens' power in the process.
   - Explicit proposals for selecting citizen representatives for the planning process.
3. Include all interested publics.
   - Staff of public agencies at the federal, state/provincial, regional, and local levels of government (as appropriate for the planning exercise).
   - Elected officials at all levels of government.
   - Private corporations and other organizations with an economic interest in the plan.
   - Public interest groups (i.e., groups formed to represent specific interests in the general public), including both high-profile leaders and the more general membership of those groups.
   - Other groups and individuals in the community, including private citizens, legal and medical professionals, and others, with a general but not necessarily economic interest in the plan.
4. Use public involvement techniques and processing for:
   - Defining the purpose of involvement.
   - Information dissemination.
   - Receiving information.
   - Two-way communication.
5. Use small-group discussion processes.
   - Brainstorming—whereby everyone contributes at the same time.
   - Delphi processes—using facilitated discussion.
   - Breakout groups.
   - Values clarification.
   - Circle processes—using iterative discussion.
   - Role play—effective for teaching conflict resolution.
   - Simulation—combines role play with random forces to create a realistic sequence of events.

These techniques provide a good way to build trust and team feeling. They permit joint analysis of complex technical issues and build mutual understanding.

Other ways to enhance the public involvement process include aspects of plan administration, data collection and analysis, and communication, in line with the following suggestions:

Plan Administration

- A single program manager and clear reporting relationships.
- Program staff who are well-informed about the project and the community, skilled in public involvement techniques, and receptive to the ideas of community representatives.
- Specialized expertise where necessary, for instance, in conflict resolution and facilitation.
- Adequate funding to achieve stated program goals for public participation.

Data Collection and Analysis

Joint collection and analysis of data on:
- Community values, systems, and interested publics.
- The proposed project and any likely impacts on community values and/or lifestyle.
- Experiences in similar projects.
- The costs of proposed measures, and possible funding sources for those measures.
- The environmental and economic impacts of proposed measures.

Communication

- Clearly written documents and legible graphics.
- Text written in semitechnical or lay language that is easily understood by all participants, and/or opportunities for technical education during the consultation period.
- Full and unrestricted access to all data, documents, and others materials for all participants, whether in electronic or hard copy form.
- Well-structured advisory and consultation groups with:
    A variety of skills represented (innovators, detail checkers, encouragers, moderators, etc.).
    A balance of activity between plan-related tasks and team-building activities. (The latter, although essential for long-term group function, should comprise no more than 40% of total activity, or the group may be unproductive.)
- Prompt, sensitive, and respectful review of citizen submissions, and thoughtful and timely bureaucrat responses to public representatives. The goal is to create an ongoing dialogue whereby ideas and suggestions from a variety of sources are welcomed and thoughtfully considered.

12.4 Summary

An integrated approach to planning, programming, and implementation-level decision making is neither as broad-based as a comprehensive planning–programming–
implementation approach nor as narrow as functional planning–programming–
implementation, because two or more themes are considered in the decision process. In
section 12.1 we addressed watershed planning as an integration of land use and water
resource planning. The integration is often between themes, such as land use and water
resources, because land use has considerable impacts on water resource quantity and
quality. In section 12.2 we addressed transportation improvement programming in
terms of a connection between land use and transportation activities. Transportation
projects are often developed in response to land use activities, because land use activities drive transportation system use. In section 12.3 we addressed wastewater facility
siting, dealing with land use, transportation, and water resources connections. An integrated approach among all three themes is common when the project size is rather large,
as in the case of the Brightwater wastewater treatment plant in Snohomish County,
Washington. The results of an integrated approach are likely to be more long-lasting,
without challenges (particularly legal challenges) to the process. However, as we have
indicated, an integrated approach is not quite a fully sustainable approach.

12.5 Review Questions

1. What is being integrated in an integrated approach to planning, programming,
   and/or implementation decision processes for resource management?

2. How would an analyst integrate functional themes for GIS data analysis in
   planning?

3. How would an analyst integrate functional themes for GIS data analysis in
   improvement programming?

4. How would an analyst integrate functional themes for GIS data analysis for project
   implementation?

5. What are the barriers to integrated water resource management described by
   Cairns (1991)? Why are they significant?

6. In which of the five phases of the integrated watershed planning process is GIS data
   analysis, including decision analysis, likely to provide decision support?

7. Why is land use so often considered in an integrated approach, whether this be for
   transportation or water resources? What would you have to do to consider land use
   in a database design for integrated planning?

8. Are the steps within an integrated approach to watershed planning useful for understanding transportation planning? Why or why not?

9. What is the most likely theme to integrate within a transportation improvement programming decision situation? How would this component in the representation model influence the subsequent (i.e., process, scenario, change, and impact) models?

10. How does a transportation adequacy measure help King County planners implement concurrency management as part of a growth management program?

11. Why is problem awareness such an important step in a public participation plan, and how would change awareness fit into this process?

12. What is a ladder of citizen participation in relation to planning processes?

13. What is meant by *citizen empowerment* in relation to planning processes?

CHAPTER 13

Linking Analyses
across Decision Situation Processes

After exploring integrated analysis among functional themes in the previous chapter, we extend the idea of integrated analysis across decision situation processes in this chapter. More efficient, effective, and equitable improvements in community quality of life must consider how plans, the budgets to implement those plans, and the implementation of projects funded by those budgets are connected to each other. Organizations across the world understand that there are links, but how to implement them is the real challenge. One way to implement the linkage is through database integration.

This chapter focuses on links among databases and analyses, highlighting much of the motivation for this book. That motivation stems from a recognized need to answer two of the seven core research questions developed as part of a National Research Council (1999) report titled *Our Common Journey*, summarized by the authors (Clark and Dickson 2003; Kates et al. 2001) in articles titled "Sustainability Science," and posted on the website for a sustainability science and technology forum (Bolin et al. 2000):

6. How can today's operational systems for monitoring and reporting on environmental and social conditions be integrated or extended to provide more useful guidance for efforts to navigate a transition toward sustainability?
7. How can today's relatively independent activities of research planning, monitoring, assessment, and decision support be better integrated into systems for adaptive management and societal learning?

In line with this motivation, as we mentioned in Chapter 1, and from a perspective on integrated watershed management in Chapter 12, Heathcote (1998 p. 391) sees a relationship among plans, programs, and implementation level management. She, like many others, recognizes that

- Plans are developed to guide programs.
- Programs are developed to match projects to social, economic, and environmental conditions in the world (i.e., what is needed, what can be done about water resource impairments).
- Projects are proposed fixes to conditions that are causing those impairments.

Commonly, different units within organizations and different organizations altogether are responsible for the planning, programming, and implementation of decision processes. Without first getting into the specific character of decision processes (as we do in sections 13.1 and 13.2), we can generalize about the distribution of effort of what might normally be used in the different phases of modeling in planning-, programming-, and implementation-level analyses (see Table 13.1). This assumes that some organizational context will be relevant to multiple people participating in the work activity.

We can compare and contrast the situations in terms of the six phases of modeling introduced in previous chapters (see Table 13.1). Assume that you have "six units of effort" to budget and expend; how would you distribute the units across each of the decision situations to know (hence, to be effective) what to do in a decision situation? Each entry in the table is the number of units that is practically adequate at that phase. By way of example in Table 13.1, we show that there is seldom enough time/ resources to do everything one would like to do, so one must make tough choices about how to allocate decision resources. As such, the effort to carry out a GIS project associated with project implementation is considerable. So information generated in the planning and the programming phases of decision making can be quite useful at the start and be carried forward for elaboration in the decision situation for project implementation.

The two sections that follow describe the link between planning- and programming-level decisions (in section 13.1) and the link between programming- and implementation-level decision processes (in section 13.2).

TABLE 13.1. Comparing Planning-, Programming-, and Implementation-Level GIS Work in Terms of Generalized Units of Effort Involved (1–3 Units) for Each Phase

Phase in the Steinitz framework	Planning	Programming	Implementation
Representation model	2	2	2
Process model	1	1	2
Evaluation model	1	1	2
Change model	1	2	2
Impact model	2	1	2
Decision model	2	3	2

13.1 Linked Analysis for Planning-Level and Programming-Level Decision Situations

Plans are developed to guide growth. Programming is conducted to finance the build-out of a plan. A linkage between planning and programming promotes a sustainable development perspective, like that of the Office of Sustainability and Environment at the City of Seattle (see Figure 13.1).

Figure 13.1 provides an overall sense of comprehensive plan linkages to city business work (capital improvement) programs. The four-step process is based on an approach to total quality management (TQM). As depicted here, the TQM process has been articulated in terms of sustainability decision modeling as the core activity (Brassard 1989).

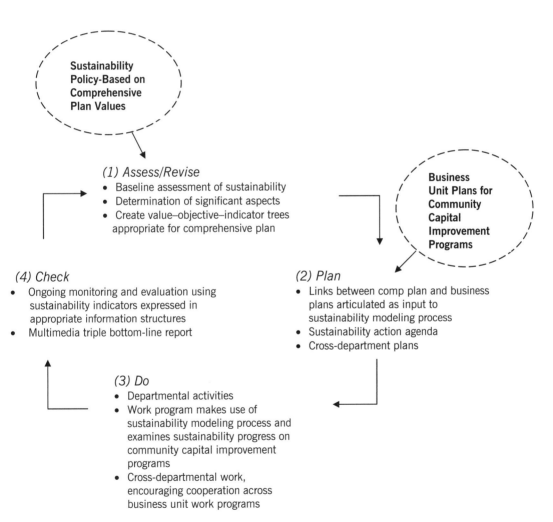

FIGURE 13.1. City of Seattle: sustainability management system model (draft).

Comprehensive plan elements in step 1 of Figure 13.1 are used to assess/revise the goals of a TQM-based sustainability process. The business programs are budgeted in step 2 to achieve these goals in step 1. A link is established between comprehensive plan goals and business program activity to set a plan of action. In step 3, sustainability decision modeling is performed to better understand how work programs can be implemented to carry out those goals. These models can follow the process outlined in Table 13.1. In step 4, monitoring and review of the sustainability modeling provide feedback to a new cycle feeding into step 1. A multimedia report provides a pleasing visual presentation of the results that is more likely to be understandable to people with diverse backgrounds. The triple-bottom-line report provides information on the social, economic, and ecological conditions of the community activities under consideration. The overall process is iterative in character, moving through steps 1–4 as a cycle, because linear processes seldom ever achieve goals without reasonable feedback. A TQM approach to decision making is a natural fit to sustainability modeling, because feedback is a very important process in both. Feedback from project outcomes to help steer implementation of plans through programs is a similar approach to business management with a TQM approach.

Generally, the goal of a linked planning and programming decision process is to ensure consistent treatment of medium- and long-term options, so that a program implements a plan. Linking comprehensive plans to capital improvement programs is an important challenge facing land-, transportation-, and water resource–oriented organizations. Of the three functional domains, the link between land use plans and programs is less clear, because land use programs are a fairly dispersed activity across society. The link between transportation plans and transportation capital improvement programs is clearer, but major challenges still exist because of the pervasive character of transportation improvement. It is perhaps most clear with water resources, in that public agencies have considerable authority over most of what happens for drinking water and wastewater, although there are many external effects.

Plans in regard to transportation were made for a long time without any "explicit" link to capital improvement programming. That is now changing. In the context of transportation, the need for this linkage is described by Younger and O'Neill (1998). A research project across state Departments of Transportation focused on the link between planning and programming (Cambridge Systematics 2006). The research uncovered several successful practices that enhance the connection between plans and programs. The research report describes insights into related aspects of the practice, including how to scope the statewide transportation plan, and meaningful approaches to public involvement that can foster a better linkage among plans and programs. Within the central Puget Sound region, both the City of Seattle and King County are using "transportation strategic plans" to bridge the gap between Comprehensive Plan transportation elements (a 20-year perspective) and short-term capital improvement programming (a 2-year perspective revolving over 6 years). This intermediate (10-year or so) perspective provides planners, citizens, and elected officials with a more effective temporal transition for decision making.

The first step in the work plan is data model integration. As such, a major challenge in linking planning with programming is their projects are conceptualized differently given the spatial and temporal scales involved. The different project conceptualizations mean that from an information perspective, plans and programs are loosely coupled at the current time, but from an information technology perspective, they are even less coupled. Working on the information technology coupling would enhance the development of visualizations that foster "shared understanding." A clearer conceptualization of the similarities and differences (as in a conceptual database design) would strengthen the link between the planning endeavor and the programming endeavor. Plans would have to take on grounded direction, and programs would obtain a longer-term justification, with some flexibility for direction.

The Puget Sound Regional Council has been addressing the concern for decision situation integration at the level of GIS database design for the Metropolitan Transportation Plan and Transportation Improvement Program (TIP). Unfortunately, there is not yet a coordinated effort across the three jurisdictional scales to address the transportation plan–program link. That "institutional concern" is one of the fundamental issues in developing successful GIS, particularly when the topic is growth management. However, GIS data framework efforts at understanding comprehensive approaches to transportation data organization are under way in Washington State (WSDOT 2006). Such efforts are being spurred on by continued work in implementing the National Spatial Data Infrastructure under the coordination of the Federal Geographic Data Committee (FGDC), with direction from the FGDC Secretariat at the U.S. Geological Survey (USGS).

Many transportation organizations are actively working on more comprehensive transportation database design projects. Most agencies recognize that the "map graphic" era of GIS is past, and that database-driven approaches to transportation information provide a much sounder perspective for the short, medium, and long term. Much of what is driving that perspective is an interest in integrating planning and programming projects.

Representing transportation planning projects and improvement projects in a GIS is actually more challenging than just placing point, line, or polygon geometries in a database, and associated symbols on a map. Planning projects are spatially abstract, whereas improvement projects tend to have multiple temporal dimensions. Planning projects come in many different categories in which multiple geometries are needed for any single category. In addition, a temporal dimension for scoping, design, and building applies to each of the different spatial geometries for improvement projects. A robust conceptual design would have to take these nuances into consideration.

Because object-based data models are now coming to widespread commercial fruition, representing improvement projects in GIS is possible; however, the conceptual database models for such applications are still crude, because no major implementations for multimodal, intermodal transportation planning and programming databases have been published in publicly available literature. The question is when, not whether, both planning projects and the improvement programming projects will be expressed in a conceptual design, a linking of plans and programs that can be supported through integrating the database designs. Because projects take on so many different dimensions, this integration would be a challenge, but a GIS database integration approach can be used to sort through the differences and the similarities (Nyerges 1989a, 1989b).

As described in Chapter 5, a conceptual schema specifies the content of a database in terms of the semantics and structure of metadata rather than data. A structural interpretation involves a description of what data classes exist, specified in terms of entity classes and attributes, and the relationships among them. A semantic interpretation involves the meaning of the data as supplied by both the data dictionary and the relationship among classes.

Nyerges (1989a) suggests that the conceptual schema integration problem can be carried out as a six-step process within a broad context of information integration for GIS (Nyerges 1989b). A general process of integration analysis is the same whether it applies to data transfer to develop geographic databases, multipurpose land information system database design, or federated database design in a land information system network. The following overview presents the steps outlined in Elmasri, Larson, and Navathe (1987), except that steps 2a and 2b in the original sequence are now steps 2 and 3. These two steps are thought to have sufficiently significant differences in the activities involved that they are separated herein.

- *Step 1: Conversion into conceptual schemas.* Convert implementation schemas into conceptual schemas using an entity–relationship (ER) or extended entity–relationship (EER) data model. The conceptual schema should include all thematic, locational, and temporal data description representations. Elmasri et al. (1987) refer to these resultant schemas as *component schemas*. In the case of spatial data transfer, the schema of the receiving system is called the *target component schema*, and the schemas of the sending systems are called *source component schemas*.

If a target schema does not exist, as may be the case with development of a new database using spatial data transfer, then a target schema should be specified before proceeding. Specify the entity types, attributes, and relationships for the target schema, as in a normal database design process. Use source schemas to determine whether definitions are available for inclusion in a target conceptual schema. Document new definitions, as well as any changed definitions, in a target data dictionary to clarify the meaning of the target schema.

If several databases are to be integrated to form a multipurpose land database or a federated database, then one schema is selected as the target and others are assumed to be sources. Alternatively, in an environment beset by the question of whose schema to use, select the entity types and attributes from each schema and label this new schema the target. This latter process, although requiring more time, may be the best compromise when facing with institutional barriers.

- *Step 2: Component schema analysis and modification.* Some constructs may need to be converted from one type to another. An attribute in one schema may be an entity type in another schema. An entity type in one schema may be a relationship in another (Kent 1984). Identifying these similarities and differences is not always easy. A review of data dictionaries can be of significant help.

- *Step 3: Identification and analysis of equivalence.* Correspondences among attributes, entity types, and relationship types are identified. Identify similar attribute field names using criteria such as the role of the attribute, whether it is a primary or secondary key, the range of domain values, or data quality, and so forth. Use the attribute similarities to assert equivalences. Use the attribute equivalences to assert entity-type equivalences. Assess the results of the comparisons and iterate, if needed. Entity types are then grouped. A group, called a *cluster*, represents a similar or related concept, as indicated by the bundle of attributes. A cluster may contain anywhere from one to all of the entity types, depending on the similarity.

- *Step 4: Integration of entity types.* Clusters of entity types are used as a guide for determining which attributes to retain and which to eliminate. Representative entity types from each cluster are chosen to retain selected attributes.

- *Step 5: Integration of relationship types and entity types created from data abstractions.* Relationships are integrated after the entity types have been integrated. Specify attribute equivalences in terms of pairs of entity type names, assert the relationship type equivalences in a type similarity matrix, and specify the relationship integration assertions to be used in relationship clusters.

- *Step 6: Creating the integrated schema.* Carry out the integration by selecting the entity type, or types, from a cluster that "best represents" the cluster identified in step 4. The same applies for relationship types in step 5. Eliminate the redundant descriptions, and append the unique descriptions to one another through explicit relationships. The result is the integrated schema.

These steps are appropriate no matter what the schema integration task. The details, with an example for land parcel database integration, can be found in Nyerges (1989a), because there is not sufficient space here. A GIS analyst could make use of any available computer-aided software engineering tool, including the Microsoft Office Visio software, which provides an ER diagramming capability to support the schema analysis process.

13.2 Linked Analysis for Programming-Level and Implementation-Level Decision Situations

In light of the previous section, improvement programming analysis is about selecting projects to be budgeted in the current round of consideration based on existing conditions. Projects can be at various phases of completion, from "not yet started and we need to scope" through "let's design the project based on the scoping effort," all the way to "build-out the project to completion according to current designs." The basis of the comparison involves a trade-off of "impacts" (i.e., what society would most or least prefer in terms of positive or negative impacts from the "package of changes"). The package is sometimes referred to as a *scenario*; that is, under certain assumptions of policy conditions, such as availability of funds, regulation of pollution levels, tax breaks, and accessibility to resource, what set of projects would best be able to address the objectives of "improved community well-being?"

An implementation-level analysis can involve a detailed investigation using several of the models associated with the Steinitz framework, but the effort for the most part goes into "impact assessment," as in social, economic, or environmental impact. A sustainability impact assessment would involve the interconnection of social, economic, and environmental conditions (Devuyst 2001). An impact assessment, whether it be social, economic, or environmental sustainability, is a detailed characterization of changes that would "likely" take place given the particular project undertaken. Devuyst has described linking impact assessments across policy, plans, programs, and projects as a process called *tiering*. The strategic impact assessment takes these linkages into consideration.

One of the most underresearched and least understood aspects of the relationship between programming-level and implementation-level analysis is the influence of "cumulative impacts." This is difficult in a sustainability impact assessment given all the functional domains that are to be treated. However, it is even more difficult when we talk about the need for impact information at the programming level, when, indeed it is the implementation level that provides this kind of information.

The goal of the linked programming- and implementation-level decision problem is to ensure consistent treatment of options in the medium term for financing and in the short-term development of the project. The data model provides the structural representation of the problem. The analysis makes use of that structural data representation to construct information representations for decision support.

As mentioned earlier, and as Table 13.1 shows, a representation model is the first phase in understanding what to do in the decision situation, whether it involves planning-, programming-, and/or implementation-level work. A database model representation for improvement programming within a growth management planning context is motivated by the policies of that context; thus, the goals, objectives, and criteria categories for inclusion reflect those policies. Organizations return to the law and their plans to guide the programming effort. In an implementation-level effort, the guideline for developing databases comes in part from policy but mostly from what is considered "best practices" knowledge. Best practices knowledge develops over a period of time through various studies about scientific and engineering impacts. It might also develop from organizational knowledge over a period of time about "what works." Best practices knowledge informs the process of implementation-level work but does not "drive" it. Organizational policies drive, laws and regulations constrain, and best practices inform the development of impact studies.

Remember that there are three phases of improvement programming—scoping, designing, building. Each phase requires slightly more information about a project to understand better the actual impacts of the improvement project. In a sense, there is an adaptive management process at hand; that is, what is known about "impacts" at the current time is fed into the programming-level work process. However, given the size of some projects, the funding amount and sources involved (commonly, large projects are defined in part by how much money is needed, and large sums of money often come from federal sources), and the anticipated impacts, an "assessment" is indicated. This might be a social assessment or an environmental assessment.

Thus, the link between programming-focused and project-focused work might best be understood in terms of the "impact model" phase of modeling, but remember that the link with decision situation work really involves all phases. However, to be practical, let us focus on the impact modeling.

At the implementation level, the analysis of impacts is about understanding what they are. We do consider alternatives of a project for what might result by selection of one alternative, but those alternatives are commonly about the same place—not different places. Thus, the database model is often more spatially and temporally constrained than the programming-level consideration. As such, the database representation tends to more detailed. So whatever database is considered for the program-level impact work with regard to a particular project, that same database design should feed into the implementation-level impact work. Once again, conceptual database model (schema) integration, as discussed in the previous section and in detail in (Nyerges 1989a), would apply as the analytic process. However, in this case we are considering schemas across decision processes rather than across functional themes.

The difference between the database models of the programming-level and the implementation-level work can be dramatic. Table 13.2 presents a collection of social and economic impacts components and indicator (criteria) variables used in social impact analysis by the U.S. Army Corps of Engineers (2009) for water projects by the district of Walla Walla, Washington. These social impacts might be considered in an implementation-level impact assessment as part of integrated watershed management. It is important to note that the focus of social impacts is on social conditions in relation to land use, transportation, water resources, and other concerns.

Table 13.3 presents a framework for an environmental impact assessment that uses an extensive list of characteristics, conditions, and factors about the environment to be set in relation to proposed actions that might cause environmental impact. If one arrays the characteristics, conditions, and factors along one axis of a matrix and the actions along another, then one recreates the Leopold Environmental Matrix in Table 13.3. The cells of the matrix can be filled in with the magnitude and importance of the impacts. For any particular project situation, by drawing a diagonal line in each cell of the matrix, one can list the anticipated magnitude of impact in the upper portion of the cell, then importance of that impact (to qualify magnitude) to the situation. Obviously, only in a computer environment would this be feasible.

There are at least a dozen ways to conduct an environmental assessment (Vanclay and Bronstein 1995). However, a few key types of EA provide an understanding of the difference between a programming-level and an implementation-level assessment; Sadler (1996) identifies, and Heathcote (1998 p. 331) summarizes, four key types of

TABLE 13.2. Sample Categories for Social Impact Assessment

Indicators/impact measure	Evaluation criteria
Residential Rate Increases	Residential rate increase > 5 percent Residential rate increase <5percent
Rate Employment Impacts	Decrease in employment > 1 percent Decrease in employment <1 percent
Power Provide Rate Risk	Public-owned utility Investor-owned utility
Fixed Income Ratepayers	Poverty rate > 10 percent of all families Poverty rate <10 percent of all families
New Power Plant Operation	Increase in employment > 1 percent Increase in employment <1 percent
New Plant Construction	Increase in regional employment > 5% Increase in regional employment <5%
Nonfishing River Recreation	Increase in employment > 1 percent Increase in employment <1 percent Short-term displacement Short-term crowding
Anadromous Fishing Recreation	Increase in employment > 1 percent Increase in employment <1 percent Short-term displacement Short-term crowding Local fishing opportunities
Site Access	Decrease in site access > 25 percent Decrease in site access <25 percent
Site Services	Decrease in site services > 25 percent Decrease in site services <25 percent
Elderly Recreationists	Over 65 years > 20 percent Over 65 years <20 percent
Social Cohesion	Increased social cohesion Decreased social cohesion
Recovery Uncertainty/Risk	Lower uncertainty of salmon recovery Higher uncertainty of salmon recovery
Business Uncertainty/Risk	Lower economic uncertainty/risk Higher economic uncertainty/risk
Extinction Risk/Existence Value	Higher extinction risk Lower extinction risk
Population Impacts	Decrease in population > 5% Decrease in population <5% Increase in population > 5% Increase in population <5%

(cont.)

TABLE 13.2. (cont.)

Indicators/impact measure	Evaluation criteria
Total Long-Term Employment	Employment losses > 5 percent Employment losses < 5 percent Increase net employment > 1% Increase net employment < 1% Decrease net employment > 1% Decrease net employment < 1%
Total Short-Term Employment	Increase in employment > 5 percent Increase in employment < 5 percent
Total Subregional Employment	Increase net employment > 1% Increase net employment < 1% Decrease net employment > 1% Decrease net employment < 1%
Aesthetics	ST: Exposed shoreline LT: Revegetated shoreline

Note. From U.S. Army Corp of Engineers (2009).

environmental assessment, although, given the wording, we should generalize this to "assessment":

1. *Strategic environmental assessment (SEA)*—process of prior examination and appraisal of policies, plans, programs and other higher-level or preimplementation initiatives.
2. *Environmental assessment (EA)*—systematic process of evaluating and documenting information on the potential, capacities, and functions of natural systems and resources to facilitating sustainable development planning and decision making in general, and to anticipate and manage the adverse effects and consequences of proposed undertakings in particular.
3. *Environmental impact assessment (EIA)*—a process of identifying, predicting, evaluating, and mitigating the biophysical, social, and other relevant effects of proposed projects and physical activities prior to making major decisions and commitments.
4. *Social impact assessment (SIA)*—process of estimating the likely social consequences that follow from specific key policy and government proposals, particularly in the context of national EA requirements.

Within the context of Heathcote's (1998) discussion of integrated watershed management, environmental assessment often includes three phases: preliminary (simple) assessment, detailed assessment, and follow-up. Thus, these four types of assessment are in some sense a matter of "workflow details," which is how we differentiate between programming-level and implementation-level assessment. Preliminary assessment is used to determine whether a project is covered by EA legislation or policy, whether an EIS is required, the necessary nature and extent of the EA process, and scoping.

TABLE 13.3. Elements of the Leopold Environmental Matrix: Described in Terms of Characteristics, Conditions, and Factors as Well as Proposed Actions

I. Existing characteristics, conditions, and factors of the environment

A. Physical and chemical characteristics

1. Earth
   a. Mineral resources
   b. Construction material
   c. Soils
   d. Land form
   e. Force fields and background radiation
   f. Unique physical features

2. Water
   a. Surface
   b. Ocean
   c. Underground
   d. Quality
   e. Temperature
   f. Recharge
   g. Snow, ice and permafrost

3. Atmosphere
   a. Quality (gases, particulates)
   b. Climate (micro, macro)
   c. Temperature

4. Processes
   a. Floods
   b. Erosion
   c. Deposition (sedimentation, precipitation)
   d. Solution
   e. Sorption (ion exchange, complexing)
   f. Compaction and settling
   g. Stability (slides, slumps)
   h. Stress–strain (earthquake)
   i. Air movements

B. Biological conditions

1. Flora
   a. Trees
   b. Shrubs
   c. Grass
   d. Crops
   e. Microflora
   f. Aquatic plants
   g. Endangered species
   h. Barriers
   i. Corridors

2. Fauna
   a. Birds
   b. Land animals, including reptiles
   c. Fish and shellfish
   d. Benthic organisms
   e. Insects
   f. Microfauna
   g. Endangered species
   h. Barriers
   i. Corridors

C. Cultural factors

1. Land use
   a. Wilderness and open spaces
   b. Wetlands
   c. Forestry
   d. Grazing
   e. Agriculture
   f. Residential
   g. Commercial
   h. Industrial
   i. Mining and quarrying

2. Recreation
   a. Hunting
   b. Fishing
   c. Boating
   d. Swimming
   e. Camping and hiking
   f. Picnicking
   g. Resorts

(cont.)

TABLE 13.3. *(cont.)*

3. *Aesthetic and human interest*
   a. Scenic views and vistas
   b. Wilderness qualities
   c. Open space qualities
   d. Landscape design
   e. Unique physical features
   f. Parks and reserves
   g. Monuments
   h. Rare and unique species or ecosystems
   i. Historical or archeological sites and objects
   j. Presence of misfits

4. *Cultural status*
   a. Cultural patterns (lifestyle)
   b. Health and safety
   c. Employment
   d. Population density

5. *Man-made facilities and activities*
   a. Structures
   b. Transportation network (movement, access)
   c. Utility networks
   d. Waste disposal
   e. Barriers
   f. Corridors

D. Ecological relationships such as:

a. Salinization of water resources
b. Eutrophication
c. Disease–insect vectors
d. Food chains

e. Salinization of surficial material
f. Brush encroachment
g. Other

Others

II. **Proposed actions that may cause environmental impact**

A. Modification of regime

a. Exotic flora and fauna introduction
b. Biological controls
c. Modification of habitat
d. Alteration of ground cover
e. Alteration of groundwater hydrology
f. Alteration of drainage
g. River control and flow modification

h. Canalization
i. Irrigation
j. Weather modification
k. Burning
l. Surface or paving
m. Noise and vibration

B. Land transformation and construction

a. Urbanization
b. Industrial sites and buildings
c. Airports
d. Highways and bridges
e. Roads and trails
f. Railroads
g. Cables and lifts
h. Transmission lines, pipelines and corridors
i. Barriers including fencing
j. Channel dredging and straightening

k. Channel revetments
l. Canals
m. Dams and impoundments
n. Piers, seawalls, marinas, and sea terminals
o. Offshore structures
p. Recreational structures
q. Blasting and drilling
r. Cut and fill
s. Tunnels and underground structures

(cont.)

TABLE 13.3.   *(cont.)*

C. Resource extraction

a. Blasting and drilling
b. Surface excavation
c. Subsurface excavation and retorting
d. Well drilling and fluid removal

e. Dredging
f. Clear-cutting and other lumbering
g. Commercial fishing and hunting

D. Processing

a. Farming
b. Ranching and grazing
c. Feed lots
d. Dairying
e. Energy generation
f. Mineral processing
g. Metallurgical industry
h. Chemical industry

i. Textile industry
j. Automobile and aircraft
k. Oil refining
l. Food
m. Lumbering
n. Pulp and paper
o. Product storage

E. Land alteration

a. Erosion control and terracing
b. Mine sealing and waste control
c. Strip mining rehabilitation

d. Landscaping
e. Harbor dredging
f. Marsh fill and drainage

F. Resource renewal

a. Reforestation
b. Wildlife stocking and management

G. Changes in traffic

a. Railway
b. Automobile
c. Trucking
d. Shipping
e. Aircraft
f. River and canal traffic

g. Pleasure boating
h. Trails
i. Cables and lifts
j. Communication
k. Pipeline

H. Waste emplacement and treatment

a. Ocean dumping
b. Landfill
c. Emplacement of tailings, spoil and overburden
d. Undergoing storage
e. Junk disposal
f. Oil well flooding
g. Deep well emplacement
h. Cooling water discharge
i. Municipal waste discharge, including spray
   irrigation

j. Liquid effluent discharge
k. Stabilization and oxidation ponds
l. Septic tanks, commercial and domestic
m. Stack and exhaust emission
n. Spent lubricants
o. Ground water recharge
p. Fertilization application
q. Waste recycling

(cont.)

TABLE 13.3.   *(cont.)*

	I. Chemical treatment	
a. Fertilization	d. Weed control
b. Chemical deicing of highways, etc.	e. Insect control
c. Chemical stabilization of soil	

J. Accidents

a. Explosions
b. Spills and leaks
c. Operational failure

Others

Note. After Munn (1975).

Detailed assessment includes analysis of impacts and mitigation necessary for the "do nothing" option. Follow-up includes monitoring and auditing functions to determine the actual impacts of the project and to ensure that necessary mitigation measures are in place. We can further characterize the difference between a programming-level and an implementation-level assessment using Steinitz's six phases of modeling in Table 13.1. A programming-level analysis performs a simple assessment. A simple assessment focuses on phases 1, 3, 4, and 6 (Steinitz et al. 2003) of modeling (values describe and evaluate options that feed the decision model in phase 6).

Thus, the programming-level assessment is like a simple EA, as presented in Chapter 11. The implementation-level EA (or EIS) is like the detailed assessment presented in Chapter 11. However, *like* is an important word. Legally, according to state and federal law, an EIS is performed only when (1) state money or federal money, respectively, is involved, and (2) a full EA is determined to be necessary because the identified impacts are "significant." When a simple EA is performed, the phrase "determination of nonsignificance" is used to describe the resultant nature of impacts.

The full impact assessment of natural and/or human activities on the environment require extensive knowledge of direct or indirect effects of different factors, and possible consequences. Besides the great possibilities of available computer-based systems to simulate, combine, and interconnect various data and processes into integrated systems, and of new information and communication technologies, there are still many problems that require further research to improve environmental impact assessment and decision-making processes. New research should concentrate on application of the innovative techniques and technologies for acquisition and management of field and experimental data; enhancement of knowledge of physical, chemical, biological, and ecological processes; development of new methodologies and modeling systems for the simulation of complex processes; and development of more flexible and user-friendly tools for decision-making processes. The wide spectrum of environmental impacts, different issues and processes, and their complexity and interdependence makes environmental impact assessment and resulting decision-making processes very difficult to perform.

13.3 Challenges in Linked Analysis: The Sustainability Analysis Challenge

As we mentioned in Chapter 1, integrated management of resources is not easy, whether integration occurs across functional themes, as in Chapter 12, or across decision situation processes. Much of the challenge is social rather than technical in nature, particularly from an interorganizational perspective. Most of the challenges involve *institutional arrangements*, which have been defined in various ways, as reviewed by Mitchell (1990). He enumerates seven dimensions of institutional arrangements: (1) legislation and regulations, (2) policies and guidelines, (3) administrative structures, (4) economic and financial relationships, (5) political structures and processes, (6) historical and traditional customs and values, and (7) key participants or actors. To gain insight into institutional arrangements, Ingram, Mann, Weatherford, and Cortner (1984) urge analysts to assemble information about the participants and their stakes, the resources to which they have access in pursuing their interests (legal rules and arrangements, economic power, prevailing values and public opinion, technical expertise and control of information, control of organizational and administrative mechanisms, political resources), and the biases inherent in alternative decision-making structures. Assembling such information is what decision situation assessment is meant to do. The previous section points out how to do that at different levels of detail. GIS learners, regardless of their level of technical expertise, should take time to appreciate such arrangements.

From the seven dimensions of institutional arrangements, Mitchell (1990) synthesizes six types of leverage points that can foster integrated management: (1) context, (2) legitimization, (3) functions, (4) structures, (5) processes and mechanisms, and (6) organizational culture and participant attitudes. The types of leverage points can be considered a framework through which to improve integration management (i.e., in many instances the flow of information among organizations to improve the effectiveness of information). Although each type of leverage point is introduced in sequence, the types of leverage points influence each other in various ways. Nonetheless, Mitchell's sequencing is intentional. Each leverage type represents a necessary but not sufficient condition for achieving integration management. In different situations, the relative importance of each leverage point could vary. Thus, the institutional framework is meant to be opportunistic in the sense that it identifies points at which leverage may be exerted to improve integration. When we combine the institutional framework with the decision situation assessment framework (Chapter 3) we provide ourselves with a comprehensive way of understanding how to affect information integration management. Below we outline the six types of leverage points of the institutional framework.

1. *Context*. To understand success or failure of integrative efforts with regard to context, an analyst should have a solid understanding of at least the following four elements (Mitchell 1990):

• *State of the natural environment* must be examined to determine whether issues or problems associated with natural systems (e.g., pollution, drought, flood, land degradation) are stimuli for action (the impairment mentioned by Heathcote [1998]).

• Prevailing *ideologies* need to be identified, because these influence the choice of goals, objectives and strategies.

• *Economic conditions* often shape people's willingness to undertake new ventures. In difficult economic times, governments are less likely to take on new initiatives.

• *Legal, administrative, and financial arrangements* provide both opportunities and constraints.

2. *Legitimization*. As boundary effects will inevitably exist, it is essential to identify the following elements (Mitchell 1990).

• *Goals and objectives* of pertinent agencies, detailed to an appropriate level.

• *Responsibility, power, or authority* of agencies to address/intervene on a problem.

• *Rules for intervention and arbitration* by higher level authorities when conflicts arise that cannot be resolved by participants who are directly involved.

3. *Functions*. Generic and substantive management functions should be linked explicitly to legitimization and to structures (Mitchell 1990).

• *Generic functions* involve (a) data collection, (b) planning, (c) regulation, (d) development, (e) monitoring, and (f) enforcement.

• *Substantive functions* are specific to a sector or resource, and include (a) supply, (b) sewage treatment, and (c) pollution control. Substantive functions concerning the linkages between land and water include (a) floodplain management, (b) erosion control, (c) drainage, and (d) wetlands. At a level that addresses linkages among water, environment, and economy, the substantive functions include (a) navigation, (b) hydroelectricity production, (c) recreation and tourism, and (d) industrial production. Comparable substantive management functions can be identified for other resources, such as agriculture, forestry and wildlife. The purpose should be to align functions to various scales as effectively as possible.

• *Scale is a key concern* when identifying which management function to associate with which scale (local, state, federal). Some functions lend themselves to the local levels (e.g., water supply; maybe this is why there is a problem in many areas), whereas others are more appropriate at the state or federal level (pollution regulation). A guiding principle is that functions should be allocated whenever possible to the scale that has the closet relationship to people receiving the service or product.

- *Different mixes of scale, generic, and substantive functions* are appropriate in different situations. A basic concern should be to examine the alignment of these features in a systematic manner, and not to assume that the pattern that has evolved over time is necessarily either the most systematic or sensible for current conditions. It may well be that the alignment made excellent sense at the time the decisions were made, but in many instances those decisions were made decades ago, when the context and conditions were significantly different.

4. *Structures.* Structures are organizational and interorganizational relationships devised to carry out functions (Mitchell 1990). When selecting structures, several issues should be considered.

- The match between functions and structural form is often imperfect, because organizational management often takes considerable time to realign workflow processes.

- Regardless of the structure chosen, boundary problems between organizational responsibility and action emerge. The nature of the structure determines whether boundary problems occur among units within a larger agency, or among a number of small, specialized agencies.

- Various permutations and combinations may be constructed, and it is misguided to think of organizational structuring in either–or terms.

- A single structure is not likely to handle all aspects of a management problem. As generic and substantive functions are allocated to different scales, it may turn out that different structures are appropriate for different scales. Thus, one should consider an array of effective interorganizational structures for different scales of watershed management.

5. *Mechanisms and processes.* Because legitimization, function, and structures are unlikely to fit together perfectly, both formal and informal mechanisms and processes are needed to facilitate bargaining, negotiating, and mediating at the boundaries of those arrangements. Whichever mix of mechanisms is chosen, a variety of processes can be drawn upon to pull together diverse viewpoints.

- *Formal mechanisms* may be used at the political and bureaucratic levels. At the political level, *interdepartmental councils* are often used to ensure that pertinent department heads meet to share views. *Select committees* often represent a cross-selection of political ideologies. At the bureaucratic level, numerous mechanisms facilitate integration and coordination. *Interdepartment committees* frequently provide a forum for exchanging information and ideas on an ongoing basis. Such structures are occasionally given the status of *commissions* that, unfortunately, are given little power. *Task forces* are usually established for a specific period of time and purpose. *Review procedures* in governmental and related agencies to circulate plans or proposals for comment.

• Many *informal mechanisms* are used to realize integrated management. Professional counterparts in different agencies can telephone one another to share information or to solicit reaction to ideas. Informal discussions before or after interdepartmental committee meetings often have more substance and accomplish more than the extended and sometimes ritualistic discussion that occurs around the interdepartmental committee table. Individuals may discuss matters of mutual concern over lunch or during work breaks. The informal processes and mechanisms do not appear on organization charts or strategic plans. At the same time, many informal actions can be invoked to block or reduce integration (e.g., factors related to culture and attitude).

• Whatever the mix of mechanisms, provisions should be made to ensure that the viewpoints of the general community and of individuals are fed into the planning and management exercises through *various processes*. In other words, some "bottom-up" elements are desirable as a counterpoint to the "top-down" orientation that characterizes resource management decision making. A bottom-up orientation leads to public participation in its various forms, and may range from consultative committees, advisory groups, and public meetings to surveys of the general public.

• *Regional planning* processes seek to incorporate a diversity of interests.

• *Cost–benefit analysis* and *environmental impact assessment* also focus attention on array of considerations.

• *Public participation* as a process serves to integrate different points of view.

• Each of these processes has relative merits. Consequently, as with the mechanisms, it is usually appropriate to determine which combination best meets the needs of the public in a given situation.

6. *Organizational culture and participant attitudes.* Integration, cooperation and coordination depend to a significant extent on the willingness of participants to make them happen (Mitchell 1990). Therefore, identifying the characteristics of the organizational culture and participants' attitudes regarding disincentives and incentives for integration becomes important. Through fuzzy legitimization, unclear functions, and cumbersome structures, an organizational culture develops that may create real barriers to integrated and cooperative effort. Organizational culture and participant attitudes can be crucial to the success of an integrated approach. Because many disincentives regarding integration exist, it is of fundamental importance that the "human dimension" be given equal consideration relative to legitimization, functions, structures, and processes/mechanisms.

• *Few explicit incentives* exist for integration and coordination. Vertical and horizontal fragmentation create an environment in which rewards usually accrue to those who concentrate on their own areas of interest. Organizational skirmishing is a function of organizational culture, which encourages most public agencies to look to their own interest first and to societal welfare second (Frost, Moore, Louis, Lundberg, and Martin

1985). Whereas society may gain through more coordination and cooperation among public agencies, some individuals may perceive themselves as becoming losers through a reduction or loss of authority. In such cases, delay, systematic disinformation, and minor sabotage often occur.

• Concerning *participants attitudes*, it is important to recall that "most decisions involving water resources are made on the basis of bargaining, negotiation, [and] compromise" (Ingram et al. 1984 p. 328). Unfortunately, resource management and in-career educational programs appear to devote little time to develop the bargaining skills of resource managers. It seems to be assumed that resource managers will gain bargaining skills through the experience of being involved in resource management. Such an approach is inefficient, because developing good bargaining skills requires the accumulation of many years of experience. An alternative approach would be to teach skills pertinent to bargaining, negotiation, and compromise.

There are several general implications for the framework leverage points presented by Mitchell (1990). If coordinated management of water and land resources is to be achieved, the scope of a holistic approach must be carefully thought through. We recommend here that a two-level approach be used. A comprehensive viewpoint is desirable at a strategic level, which implies scanning the widest possible range of issues and variables. A more focused approach should be utilized, at an operational level, however, where attention is concentrated on issues and variables that are judged to be significant. Thus, a bounded holistic perspective is achieved, which should lead to findings and recommendations that are both timely and related explicitly to the needs of managers.

Because of its broad-based nature, the institutional framework is applicable to substantive areas other than water resources management integration, such as land use, transportation, economic development, and so forth. The framework can be used in either a descriptive mode or a prescriptive mode, just like the decision situation assessment framework. In a descriptive mode, the framework helps to concentrate attention on key events, decisions, and people in a resource management situation. In a prescriptive mode, guidelines that are part of the framework provide direction for possible changes or modifications to a management situation. Recommendations should be custom-designed with regard to pertinent characteristics of a situation.

13.4 Summary

Much of the motivation for this textbook stems from a recognized need to answer two major questions about the research, development, and practice of sustainability:

1. How can today's operational systems for monitoring and reporting on environmental and social conditions be integrated or extended to provide more useful guidance for efforts to navigate a transition toward sustainability?

2. How can today's relatively independent activities of research planning, monitoring, assessment, and decision support be better integrated into systems for adaptive management and societal learning?

Making a direct link among planning, improvement programming, and project implementation decision processes can foster more sustainable decision making. Plans and improvement programs must share data, as must improvement program and implementation situations. We can move beyond the one-off, sustainable development project and institutionalize a sustainability perspective. It is much easier to change processes when information capabilities provide direct connections among the decision situations.

Progress with instituting a sustainability perspective is occurring. The idea is several years old, but it is not easy to change in practice. Local and regional governments can be among the first organizations to adopt this perspective, because they are commonly charged with regional improvement anyway. Strategic sustainable assessments are under way in a variety of places (Devuyst 2001), mostly in Europe, but the United States can catch up.

Some of the major challenges for adopting a sustainability perspective are similar to those for adopting the integrated watershed management approach set forth by Mitchell (1990) several years ago. He enumerated seven dimensions to institutional arrangements that can be used as leverage points for integration across and within organizations: (1) legislation and regulations, (2) policies and guidelines, (3) administrative structures, (4) economic and financial relationships, (5) political structures and processes, (6) historical and traditional customs and values, and (7) key participants or actors. The aspects of information integration that concern people are often more challenging than the technical aspects. Given that GIS information technology is inherently integrative in character, it appears that GIS is appropriate for the job.

13.5 Review Questions

1. In what way does sustainability provide the primary motivation for linking planning, programming, and project implementation processes?

2. How are plans, programs, and projects related?

3. What role does GIS-based modeling play in the link between planning and programming?

4. What role does GIS-based modeling play in the link between programming and implementation?

5. What is the relationship between total quality management and sustainability planning?

6. What is the first step that needs to be addressed when trying to link planning and programming processes or, alternatively, when trying to link programming and

implementation processes? How would you address that step in transportation decision processes?

7. What opportunity exists for examining cumulative impacts in regard to project implementation? Are impacts more prevalent at the programming or at the project implementation level?

8. Describe the kinds of issues associated with social impact assessment.

9. How can we differentiate between environmental assessment and environmental impact assessment?

10. What are the challenges in linked analysis in terms of the six leverage points outlined by Mitchell in regard to integrated resource management?

PART V

Concluding Perspective

CHAPTER 14

Perspectives on GIS and Sustainability Management

The chapters in this book are designed to enhance a GIS learning experience by using a decision support perspective. The material is set in a substantive context of urban–regional sustainability planning, improvement programming, and project-level decision processes, with special focus on the links among land use, transportation and water resources. Grounding the book in a substantive context in the everyday world that guides significant change within urban–regional sustainability settings is thought to inform readers about the considerable benefit of GIS use in society. This last chapter provides reflections on the current state of GIS, particularly as it relates to decision support and some of the challenges for the future of GIS use in society.

14.1 Decision Situation Assessment

The situations outlined in Chapter 1, but presented in detail in Chapters 2 and 3, set the stage for understanding how complex many of the land, transportation, and water resource decision situations can be. The decision situation assessment framework called *enhanced adaptive structuration theory* (EAST2) presented in Chapter 4 encourages us to consider the complexity of GIS use for planning, programming, and project-level decision support from a broad perspective, because many aspects of decision situations in society should be considered when putting GIS technology to use. We can adopt an approach that unpacks decision support complexity into various components, constructs, and aspects, making such complexity more understandable. In particular, we should consider the institutional influences (i.e., the laws, policies, regulations, missions, guidelines, and important research questions not yet answered, etc.) that direct industries, governments, not-for-profit organizations, groups, and individuals to consider geographic information in various ways when addressing planning, programming, and project-level issues. We need to understand the influence that various communities and publics bring to situations through their motivation to address certain concerns based on interpretations of institutional influences. Those interpretations in large part motivate certain values, goals, objectives, and criteria measurements to be used (hence, the databases to be developed and analyses to be performed) when people support planning, programming, and project-level decision tasks with GIS-based information products.

With so many opportunities to consider for GIS use, it is unfortunate that we are limited by space in this book. As with most situations with considerable opportunity, one must choose a focus and establish priorities. Although our efforts focus on land use, transportation, and water resource issues, we recognize that these issues and their interrelationships are certainly not the only issues of deep concern in urban–regional communities. Nonetheless, our issues of focus are among the most fundamental in urban–regional society as recognized and stated in growth management laws across the United States and around the world when we consider improvements to quality of life in a community. Although growth management laws exist in only a limited number of states across the United States, the connections among land use, transportation, and water resources are widespread because of the everyday behavior of people within communities all across the world. Thus, all communities everywhere have an opportunity to recognize how plans, programs, and project-level activities influence the sustainability of well-being and improving quality of life for current, as well as future, generations. This is indeed a rather broad-based policy agenda.

14.2 Growth Management Perspective

As of 2006, 11 of the 50 U.S. states have enacted growth management laws; some others are considering them, and several others are considering some form of growth management activity. Three states—Florida, New Jersey, and Oregon—have been using top-down controls (i.e., a strong state-level control to encourage development growth). Eight states—Georgia, Hawaii, Maine, Maryland, Minnesota, Rhode Island, Vermont, and Washington—use bottom-up control (i.e., stronger local-level control). A 12th state—California—is beginning to use a combination of both. Twenty-seven states have a role in growth management, but 13 states have no mandated state laws.

In "top-down planning" states, such as Oregon, goals are more specific at the state level than they are in "bottom-up planning" states, such as Washington. In top-down states, the goals are stated in such a way that all counties within the state plan in the same way. In bottom-up planning states, the goals are generalized but made specific by local jurisdiction implementation as long as the jurisdiction makes some kind of plan. Certain thresholds about development may differ from jurisdiction to jurisdiction.

Communities function within city, regional, state, national, and even global contexts of economic, social, and environmental conditions—the conditions examined in the sustainable development literature. There are always multiple scales and intertwined goals. As such, many issues are in need of coordination if growth management is going to work. Coordination across scales, meaning across boundaries of adjoining jurisdictions, is critically important to making progress with urban–regional growth management, and even more challenging for urban–regional sustainability. Several forces in democracies make growth management a challenge.

First, older democracies tend to foster a local perspective. American culture tends to reinforce the idea that public decision making should happen at the lowest possible level of government to be most meaningful. This local perspective fosters skepticism about "big government" (i.e., government that spends money to try to fix local problems).

Second, local influences tend to resist regional and state influences. Even if these influences are accepted, there is still tension that results in intergovernmental challenges.

Third, despite resistance by local interests in relation to regional and state interests, extralocal interests tend to demand attention, because boundaries are open to human and wildlife behavior. Some examples of these tendencies are as follows:

- Transportation systems require coordination at regional and state levels, and in fact, by law, for a region to receive federal dollars that coordination must be documented.
- Sewer and water systems cover multiple jurisdictions to make them more efficient.
- Watersheds cross jurisdictional boundaries. The fish pay no attention to city boundaries; hence, recently formed watershed planning councils are coordinating bodies.
- Social and economic disparities among jurisdictions threaten to disrupt regional economies unless addressed on regional basis.

These circumstances lead to the need for regional growth management, implemented through various organizations to foster more effective management. Some of the types of regional organizations that participate in growth management are the following:

- Regional planning councils or districts within states (e.g., in Washington State, regional transportation planning organizations are required to receive state funds in urban–regional areas).
- Metropolitan planning organizations for all major population centers (cities).
- Federal/state chartered commissions or authorities are charged with protecting environmentally sensitive areas.
- Regional public service authorities (e.g., in the Seattle area, the Port of Seat-

tle manages both harbors and airports), and there are regional transportation authorities as well, such as Sound Transit.

- Regional business and civic leadership groups, such as City and County Chambers of Commerce.
- Ad hoc groups, such as the Water Forum and Tri-County Water Coalition, that are each three-county organizations, but not mandated.

Based on many of these experiences, one can learn several lessons relative to effective regional growth management. The more successful organizations listed earlier are effective because they (1) have a broad constituency of interests (e.g., transportation and/ or environment across the region), (2) have clear objectives for action (e.g., addressing water pollution to clean up Lake Washington in 1970s), (3) focus on regional matters and have the ability to guide state/federal authorities to say "no" to some local proposals, (4) experience local accountability relative to regional interests (e.g., coordination of county transportation plans), and (5) foster shared regional decision making, but local government retains major responsibility to implement it.

14.3 Perspectives on Sustainability Management

It is advantageous to relate community and regional sustainability to growth management, and growth management to conventional community management, if community and regional sustainability are to make progress within current institutional contexts. Drawing growth management and sustainability views into focus, in Figure 2.3 we suggested a perspective about "community and regional sustainability" that makes use of Farrell and Hart's (1998) description of competing social, economic, and environmental objectives for communities that may or may not be considered together with carrying capacities, and Rees's (1998) discussion of the importance of generational equity in sustainable community development. The four-cell framework characterizes community and regional sustainability in terms of three levels—weak, semistrong, and strong. Competing objectives, carrying capacity, and intra- and intergenerational equity combine to form a progression of weak to strong community and regional sustainability. Growth management concerns are about competing objectives and intra- and intergenerational equity (weak and semistrong sustainability), but growth management seldom addresses social, economic, and environmental concerns simultaneously. The natural, physical, and social sciences continue to assess carrying capacity related to various social, economic, and ecological concerns that are the basis of "integrated assessment science" and considered the core of "sustainability science" (Kates et al. 2001). Sustainability assessment cuts across jurisdictional boundaries, for example, in watershed sustainability studies. Watersheds do not align themselves nicely with political jurisdictional governance—nor do the problems of sustainability.

We can use guiding principles for sustainable urban development as a framework and motivation for GIS work in an urban–regional setting. The guiding principles framework, presented by Haughton and Hunter (1994) in their text *Sustainable Cities* is a synthesis based on several authors' work.

They develop a three-tier framework of "guidance" for urban sustainable development. The first two tiers apply globally. The third tier applies to the local scene (i.e., how the local (region) community wishes to put the first two tiers to action).

First-Tier Fundamental Principles

- Intergenerational equity—same stock of resources for future.
- Social justice—intragenerational equity, satisfying all basic needs.
- Transfrontier justice—cross-boundary exportation of environmental problems.

Second-Tier Guiding Principles

- Selected principles about ecological, socioeconomic, management issues.

Third-Tier Desirable Policy Directions

- The comprehensive plan to put the first two tiers to action is left up to the local community.

The question for the reader is: What are the implications of these principles in relation to GIS work in a broad sense?

In regard to the *second-tier of principles in the "ecological" category*, for example:

- *Ecological principle 1: Prevention is better than cure.* Use environmental impact assessment to know what impacts are occurring (the same for ecological, social, and economic assessments).
- *Ecological principle 2: Nothing stands alone.* Account for the local, regional, and global implications of urban activities.

What are implications for GIS activity? Both of these principles beg the use of GIS to address sustainability assessment as a combination of ecological, social, and economic assessment, and at the same time account for the implications of activities in those assessments, that is, the influences of development activities in multiple domains. Devuyst, Hens, and De Lannoy's (2001) book *How Green Is the City?* makes reference to how GIS can assist with *sustainability assessment* as an extension of environmental assessment. An entire book can be written about the use of GIS to address Haughton and Hunter's (1994) first ecological guiding principle. The applications exist; it will take several people to follow through and show how important GIS can be to address these ecological principles. This book is but a first step, and there is much more work to be done.

In regard to the *second-tier of principles in the "social and economic" category*, for example:

- *Socioeconomic principle 2: Create new indicators for economic and environmental wealth.* Indicators need be developed at relevant spatial and temporal scales to show distributions of economic and environmental wealth (e.g., the genuine progress indicator; see *www.rprogress.org/projects/gpi*).
- *Socioeconomic principle 6: Ensure social acceptability of environmental policies.* Policies designed to improve the urban environment should not result in a net decline in the quality of life of disadvantage groups, both in cities and globally.
- *Socioeconomic principles 7: Widespread public participation.* This principle is encouraged in strategy formulation (plan making), policy implementation (capital improvement program), and project management (efficiencies and monitoring of what has been done).

What are implications for GIS activity? GIS can be used to manage data for indicator and index development not only at the national level but also at the local level (e.g., city and county indicators of well-being were addressed in various sections of the book). Indicators can be used to "ground information" in both plan making and impact assessment. Impact assessments are needed for better policy development to make plans that result in improved decision processes. Public participation supposedly adds to breadth and depth of decision processes. Such participatory processes might take slightly longer, but those processes are likely to have fewer court challenges, because they included people. This was the major reason for implementing specialized GIS software for water resource planning called WaterGroup (Jankowski et al. 2006).

In regard to the *second-tier of principles in the "management" category*, for example:

- *Management principle 1: Subsidiarity.* Responsibility for implementation and management of urban programs is at the lowest feasible and appropriate level—the local level—of government.
- *Management principle 6: Need for better availability and understanding of environmental information.* Communities and business should be informed of environmental consequences of development proposals, including cross-boundary concerns.

What are implications for GIS activity? Both management principles 1 and 6 encourage the use of GIS at the most disaggregated level of decision making possible. That encouragement parallels values underlying democratic process and economic effectiveness (but not efficiency). Such motivation is one of the major reasons why democratic countries around the world make considerable use of GIS, and particularly why the decentralized approach, common in the United States, makes it the largest user and producer of GIS information technology—a fine-grain, disaggregated perspective on the world.

When looking at these sustainability principles, we can identify connections to growth management. The challenge is simply to take a slightly broader but more fundamental perspective.

14.4 Overall Implications for GIS-Oriented Activity

How well does GIS technology address growth management and sustainability concerns, and what might we expect in the future? There is no doubt that GIS can support both growth and sustainability management at this time. However, full support for all analyses and management is not yet possible for most commercial GIS. It is a matter of marketplace recognition, which is still to come. We can characterize the capabilities of GIS to address growth and sustainability management in terms of database management, analysis, and mapping.

Integrated database management is possible. Establishing links among planning, improvement programming, and project implementation databases is certainly possible given the current technology. The challenge is to get different parts of organizations to collaborate on their database designs, because each decision situation has its own decision requirements.

The biggest obstacle for growth management is characterizing change over time, but it is possible to design databases that consider temporal dimension. In Chapter 3 we discussed a nuanced workflow process developed by Steinitz and his colleagues (2003). The modeling activity makes considerable use of GIS, but some of the workflow steps are better performed by software other than GIS. The representation model development is undoubtedly a GIS-based activity, because databases are foundational to GIS work. The process model, however, might be better implemented through other specialized software, because most GIS software has not yet been designed to address temporal data processing issues. Spatial–temporal modeling (e.g., land use change over multiple increments of time) can be done as time slices, but the analysis is actually more static, with the process shown as a visual animation. Using spatial–temporal modeling better to understand change, a field of research called "geographic dynamics," is still emerging (Yuan and Stewart Hornsby 2007). Scenario, change, and impact modeling are readily performed with GIS software, but without advanced process models the outcomes are still not as realistic as some would like. Decision modeling is still somewhat challenging; once again the algorithms are somewhat specialized, and GIS vendors have not yet fully adopted a variety of them for application.

Mapping has traditionally been a major part of GIS, so the current functionality is quite good to address growth management and sustainability management activity. Time-slice animation is possible and is undertaken quite frequently. Three-dimensional visualization software that implements smooth animation is superior, but such software does not have the analysis capability of GIS software. Moving analysis output into three-dimensional animation software is what a number of organizations do when they want to make GIS movies.

GIS, as an information technology, and particularly a decision support technology in a broad sense, will mature only if we challenge it to address complex and demanding problems. Furthermore, we will not develop a useful expertise unless we challenge ourselves to use GIS technology in complex ways. Group-based decision support of land use, transportation, and water resources within planning, programming and project-level processes is among the more complex and important topics in the 21st century.

When we broaden the topic of group decision support to public decision support, we start to address fundamental issues in the democratization of decision processes. Considerable research is under way to place GIS in participatory contexts, whether it is called participatory GIS (PGIS; Harris, Weiner, Warren, and Levin 1995; Jankowski and Nyerges 2001a) or public participation GIS (PPGIS; Nyerges 2005; Nyerges, Barndt, and Brooks 1997), or community integrated GIS (Craig et al. 2002). Regardless of the label, individuals as part of the public, and groups within the public, are often marginalized in public decision processes. When we examine the ability to give public voice a hearing in democracy, the marginalized voice is a fairly pervasive problem. Practically speaking, the general public comprises many diverse groups—even if we consider the public as *whole*. The *general public* is actually a marginalized group when it comes to participation processes, because there is no single directed voice of the public.

Despite many federal, state, and local laws that require public participation, research about local governance indicates that large-group participation in publicly oriented decisions commonly involves little *meaningful participation*, which can be defined in terms of *access to voice* (a deliberative process) and *competence of knowledge(s)* (an analytical process) that fosters *shared understanding* about values, interests, and concerns that underlie the recommendations/choices to be offered/made by those with a stake in the decision (National Research Council 1996). Meaningful participation is a hallmark of a healthy democracy, particularly a deliberative democracy in contrast to a representative (make a vote) democracy.

Deliberative democracy involves empowerment in which reasoned discussion among people promotes shared understanding on a topic, followed by consensus building. Although interest in deliberative democracy has existed for over 100 years (Gastil and Levine 2005), research and practice have blossomed only since the late 1980s. Over the past decade, hundreds of deliberative democracy events of varying sizes have occurred across the world. A synthesis of case studies appears in *The Deliberative Democracy Handbook* (Gastil and Levine 2005). Several of the chapters deal with location-based issues; thus, GIS could be useful. However, no chapters actually refer to GIS, a seeming disconnect and latent opportunity.

Research about analytical–deliberative decision processes has shown that meaningful public participation is possible, and decision outcomes are improved (National Research Council 1996). The analytical component provides technical information that ensures broad-based, competent perspectives. GIS has provided technical information in the form of maps that represent changes in landscapes. The deliberative component provides an opportunity to give voice to choices about values, alternatives, and recommendations. Unfortunately, such public participation has been expensive and time consuming, and involved small- to medium-size groups (10–15 people). Working through analytical–deliberative participation in small- to medium-size groups in face-to-face settings is a start, but scaling analytical–deliberative participation out to include large groups is a challenge, and scaling up (e.g., from local to regional domains) is also a challenge; but scaling out and up matters (Nyerges 2005).

In addition, whether groups are better supported in face-to-face settings or in asynchronous settings is still an open research question. It is often thought that face-to-face participatory settings are superior to asynchronous participatory settings. It only seems reasonable. However, Dowling and St. Louis (2000) showed that an asynchronous nominal group process was more effective than a face-to-face nominal group process, at least in a small-group setting—a challenge to anecdotal feelings about face-to-face participatory processes.

Based on the following three observations—(1) public participation is mandated by many federal, state, and local laws encouraging core democratic process; (2) the Internet is growing in popularity and access is getting better even for underrepresented groups, as reported in several studies; and (3) asynchronous, structured participation methods have been shown to be at least as good as, and in some cases are superior to face-to-face participation—perhaps an Internet platform combining GIS (i.e., data management, spatial analysis, and geovisualization) technologies, decision-modeling technology, and communications technology into a geospatial portal to support an analytical–deliberative process might be one way to foster meaningful participation in large groups, as well as hold down the cost to all who wish to participate. That rationale was the basis of the U.S. National Science Foundation–funded research activity called the Participatory GIS for Transportation (PGIST) Project (Young et al. 2007). The PGIST Project hosted a field experiment in supporting public participation in transportation decision making as a glimmer of what might be coming in web-based technologies for public decision support (*www.letsimprovetransportation.org*). Emergent societal trends suggest that more and more people do care about the sustainability of their communities. GIS can help shed light on new directions.

The integration of geographic information across those space–time decision scales can be a practical foundation, supported by methodological and theoretical foundations in GIS science, for addressing growth management and sustainability concerns in the 21st century. Web-based information technologies are developing so quickly that, clearly, GIS implemented with such technologies will likely have significant impacts on society in the future, as more people explore the advantages of spatial thinking (National Research Council 2006). How geospatial information technologies get developed, deployed, and used can be influenced by those who care enough to put forward an effort to make a difference. Individuals and communities will collectively decide.

14.5 Review Questions

1. What role does decision situation assessment play in understanding the need for GIS-based decision support capabilities?

2. What is the relationship among conventional management of community development, growth management, and sustainability management?

3. What are the regional growth management challenges?

4. What are the implications for the future of GIS applications in communities around the world when considering growth management?

5. What are the implications for the future of GIS applications in communities around the world when considering urban–regional sustainability management in terms of the first-tier principles?

6. What are the implications for the future of GIS applications in communities around the world when considering urban–regional sustainability management in terms of the second-tier principles?

7. How would you characterize the overall implications for GIS activity related to sustainability management in terms of the future developments of GIS-oriented database management, spatial analysis, and map visualization, knowing that information integration technology needs for the future will grow rather than decline?

References

3D Nature. 2006. 3d Nature Studio software description. Retrieved August 22, 2006, from *www.3dnature.com/vnsinfo.html.*

Alachua County, Florida. 2008. GIS Applications in Alachua County. Retrieved February 20, 2008, from *growth-management.alachuacounty.us/gis/gis_index.php*

Albrecht, J. 1999. Universal analytical GIS operations: A task-oriented systematization of data structure-independent GIS functionality. In M. Cgraglia and H. Onsrud (Eds.), *Geographic Information Research: Trans-Atlantic Perspectives.* London: Taylor & Francis, pp. 577–591.

American Heritage Dictionary. 2006. *American Heritage Dictionary of the English Language* (4th ed.) Retrieved November 12, 2006, from *www.bartleby.com/61.*

Andrienko, G., Andrienko, N., and Gatalsky, P. 2000. Towards exploratory visualization of spatio-temporal data, 3rd AGILE conference on Geographic Information Science, Helsinki/ Espoo, May 25–27, 2000, pp. 137–142.

Andrienko, G., Andrienko, N., and Jankowski, P. 2003. Building spatial decision support tools for individuals and groups. *Journal of Decision Systems,* 12(2), 193–208.

Arctur, D., and Zeiler, M. 2004. *Designing Geodatabases.* Redlands, CA: ESRI Press.

Arnstein, S. 1969. A ladder of citizen participation. *Journal of the American Institute of Planners,* 35, 216–224.

Barber, B. 1984. *Strong Democracy: Participatory Politics of a New Age.* Berkeley: University of California Press.

Beecher, J. A., and Shanaghan, P. E. 2002. Strategic planning framework for small water systems. In L. W. Mays (Ed.), *Urban Water Supply Handbook.* New York: McGraw-Hill, pp. 2.1–2.14.

Bolin, B., Clark, W., Corell, R., Dickson, N., Faucheux, S., Gallopín, G., et al. 2000. *Core Questions of Science and Technology for Sustainability.* Retrieved November 10, 2006, from *sustsci.aaas. org/content.html?contentid=776.*

Brail, R. K., and Klosterman, R. E. (Eds.). 2001. *Planning Support Systems: Integrating Geographic Information Systems, Models, and Visualization Tools.* Redlands, CA: ESRI Press.

Brassard, M. 1989. *The Memory Jogger Plus +: Featuring the Seven Management and Planning Tools.* Meuthen, MA: GOAL/QPC.

Brown, P. 1989. Property maps. In P. M. Brown and D. D. Moyer (Eds.), *Multipurpose Land Information Systems: The Guidebook* (Vol. 1). Washington, DC: Federal Geodetic Control Committee, pp. 5-1–5-21.

Cairns, J. 1991. Future needs. In J. Cairns and T. V. Crawford (Eds.), *Integrated Environmental Management*. Chelsea, MI: Lewis, pp. 183–191.

California Resources Agency. 2005. *A Citizen's Guide to Planning*. Retrieved January 10, 2005, from *ceres.ca.gov/planning/planning_guide/plan_index.html*.

Cambridge Systematics. 1996. *A Framework for Performance-Based Planning: Task 2 Report: National Cooperative Highway Research Program*. Washington, DC: Transportation Research Board.

Cambridge Systematics. 2006. Factors That Support the Planning–Programming Linkage, NCHRP Project 8-50. Retrieved June 11, 2008, from *www.trb.org/trbnet/projectdisplay.asp?projectid=926*.

Carver, S. J. 1991. Integrating multi-criteria evaluation with geographical information systems. *International Journal of Geographical Information Systems, 5*(3), 321–339.

Chen, P. P.-S. 1976. The entity–relationship model: Toward a unified view of data. *ACM Transactions on Database Systems, 1*(1), 9–36.

Chrisman, N. 1999a. What does GIS mean? *Transactions in GIS, 3*(2), 175–186.

Chrisman, N. 1999b. A transformational approach to GIS operations. *International Journal of Geographical Information Science, 13*(7), 617–637.

Chrisman, N. 2002. *Exploring geographic information systems* (2nd ed.). New York: Wiley.

Chrisman, N. 2005. *Charting the Unknown: How Computer Mapping at Harvard Became GIS*. Redlands, CA: ESRI Press.

City of Seattle. 2006a. U.S. Mayors Climate Protection Agreement. Retrieved July 29, 2006, from *www.seattle.gov/mayor/climate/default.htm#cities*.

City of Seattle. 2006b. Water System Plan. Retrieved June 16, 2006, from *www.seattle.gov/util/about_spu/water_system/plans/water_system_plan/index.asp*.

City of Seattle. 2006c. Drainage Program Plan. Retrieved June 16, 2006, from *www.seattle.gov/util/stellent/groups/public/@spu/@rmb/@resplan/@plancip/documents/spu_informative/2004co pre_200406021630476.pdf*.

City of Seattle. 2006d. Capital Improvement Program Adopted 2005. Retrieved September 1, 2006, from *www.seattle.gov/financedepartment/0510adoptedcip/default.htm*.

Clark, W., and Dickson, L. 2003. Sustainability Science: The Emerging Research Program. Retrieved November 10, 2006, from *ksgnotes1.harvard.edu/bcsia/sust.nsf/pubs/pub78*.

Clarke, K. C. 2003. *Getting Started with Geographic Information Systems* (4th ed.). Upper Saddle River, NJ: Prentice-Hall.

Codd, E. F. 1970. A relational data model for large shared data banks. *Communications of the ACM, 13*(6), 377–387.

Colgan, C. S. 2003. *The Changing Ocean and Coastal Economy of the United States*. A briefing paper for conference participants at the National Governor's Center for Best Practices Conference, Waves of Change: Examining the Role of States in Emerging Ocean Policy. October 22, 2003. 18 pp.

Cova, T. J., and Goodchild, M. F. 2002. Extending geographical representation to include fields of spatial objects. *International Journal of Geographical Information Science, 16*(6), 509–532.

Cowen, D. 1988. GIS versus CAD versus DBMS: What are the differences? *Photogrammetric Engineering and Remote Sensing, 54*(11), 1551–1555.

Craig, W., Harris, T., and Weiner, D. (Eds.). 2002. *Community Participation and GIS*. London: Taylor and Francis.

Crosetto, M., and Tarantola, S. 2001. Uncertainty and sensitivity analysis: Tools for GIS-based model implementation. *International Journal of Geographical Information Science*, 15(5), 415–437.

Crosset, K., Culliton, T., Wiley, P., and Goodspeed, R. 2004. *Population Trends Along the Coastal United States: 1980–2008*. Coastal Trends Report Series. Silver Spring, MD: NOAA.

DeSanctis, G., and Gallupe, R. B. 1987. A foundation for the study of group decision support systems. *Management Science*, 33, 589–609.

DeSanctis, G., and Poole, M. S. 1994. Capturing the complexity of advanced technology use: Adaptive structuration theory. *Organization Science*, 5(2), 121–147.

Devuyst, D. 2001. Linking impact assessment with sustainable development and the introduction of strategic environmental assessment. In D. Devuyst, L. Hens, and W. De Lannoy (Eds.), *How Green is the City?* New York: Columbia University Press, pp. 129–155.

Devuyst, D., Hens, L., and De Lannoy, W. (Eds.). 2001. *How Green Is the City?* New York: Columbia University Press.

Dowling, K. L., and St. Louis, R. D. 2000. Asynchronous implementation of the nominal group technique: Is it effective? *Decision Support Systems*, 29, 229–248.

Drew, C. H. 2002. *The Decision Mapping System: Promoting Transparency of Long-Term Environmental Decisions at Hanford*. Doctor of Philosophy thesis, University of Washington, Seattle.

Dzurik, A. A. 2003. *Water Resources Planning*. Lanham, MD: Rowman and Littlefield.

Eastman, J., Jin, W., Kyem, P., and Toledano, J. 1995. Raster procedures for multicriteria/multi-objective decisions. *Photogrammetric Engineering and Remote Sensing*, 61(5), 539–547.

Edwards, W., and von Winterfeldt, D. 1987. Public values in risk debates. *Risk Analysis*, 7(2), 141–158.

Egenhofer, M. J., and Franzosa, R. D. 1991. Point-set topological spatial relations. *International Journal of Geographical Information Systems*, 5(2), 161–174.

Elmasri, R., Larson, J., and Navathe, S. 1987. *Schema Integration Algorithms for Logical Database Design and Federated Databases* (Technical Report CSC-86-9:8212), Golden Valley, MN: Honeywell, Inc., Computer Sciences Center Library, MN63-CO60

Environmental Systems Research Institute. 2002. *Getting Started with ArcGIS*. Redlands, CA: ESRI Press.

Environmental Systems Research Institute. 2003. Building Geodatabases with CASE Tools. Retrieved November 17, 2006, from *support.esri.com/index.cfm?fa=knowledgebase.documentation.viewdoc&pid=43&metaid=658*.

Environmental Systems Research Institute. 2006. Data Model Gateway. Retrieved November 15, 2006, from *support.esri.com/index.cfm?fa=downloads.datamodels.gateway*.

Farrell, A., and Hart, M. 1998. What does sustainability really mean? *Environment*, 40(9), 4–13.

Ford, A. 1999. Modeling the Environment: An Introduction to Systems Dynamics Models of Environmental Systems. Washington, DC: Island Press.

Foresman, T. W. 1998. GIS early years and the threads of evolution. In T. W. Foresman (Ed.), *The History of Geographic Information Systems: Perspectives from the Pioneers*. Upper Saddle River, NJ: Prentice-Hall, pp. 3–17.

Frost, P., Moore, L., Louis, M., Lundberg, C., and Martin, J. (Eds.). 1985. *Organizational Culture*. Beverly Hills, CA: Sage.

Gastil, J., and Levine, P. (Eds.). 2005. *The Deliberative Democracy Handbook: Strategies for Effective Civic Engagement in the Twenty-First Century.* San Francisco: Jossey-Bass.

Geertman, S. C. M. 2002a. Participatory planning and GIS: A PSS to bridge the gap. *Environment and Planning B, 29,* 21–35.

Geertman, S. C. M. 2002b. Technology for planning, review of planning support systems: Integrating geographic information systems, models and visualization tools. *Journal of the American Planning Association, 68*(3), 318–319.

Geertman, S. C. M., and Stillwell, J. 2004. Planning support systems: An inventory of current practice. *Computers, Environment and Urban Systems, 28*(4), 291–310.

Georgia Quality Growth Partnership. 2006. Georgia Quality Growth website. Retrieved August 22, 2006, from *www.georgiaqualitygrowth.com/.*

Gersib, R., Aberele, B., Driscoll, L., Franklin, J., Haddaway, B., Hilliard, T., et al. 2004. Enhancing Transportation Project Delivery Through Watershed Characterization, Washington State Department of Transportation. Retrieved December 27, 2007, from *www.wsdot.wa.gov/environment/watershed/characterization.htm.*

Goodchild, M. F., Yuan, M., and Cova, T. J. 2007. Towards a general theory of geographic representation in GIS. *International Journal of Geographical Information Science, 21*(3), 239–260.

Harris, T. M., Weiner, D., Warner, T. A., and Levin, R. 1995. Pursuing social goals through participatory geographic information systems: Redressing South Africa's historical political ecology. In J. Pickles (Ed.), *Ground Truth: The Social Implications of Geographic Information Systems.* New York: Guilford Press, pp. 196–222.

Haughton, G., and Hunter, C. 1994. *Sustainable Cities.* London: Kingsley.

Heathcote, I. W. 1998. *Integrated Watershed Management: Principles and Practice.* New York: Wiley.

Hillsborough County. 2006. Capital improvement programs tracked by the project information management system. Retrieved September 1, 2006 From *www-gis.hillsboroughcounty.org/gis/pims/CIPbook.cfm.*

Hopkins, L. D. 2001. *Urban Development: The Logic of Making Plans.* Washington, DC: Island Press.

Ingram, H., Mann, D., Weatherford, G., and Cortner, H. 1984. Guidelines for improved institutional analysis in water resources planning. *Water Resources Research, 20,* 323–334.

Innes, J. E. 1996, Autumn. Planning through consensus building: A new view of the comprehensive planning ideal. *Journal of American Planning Association,* pp. 460–472.

Innes, J. E. 1998, Winter. Information in communicative planning. *Journal of American Planning Association,* pp. 52–63.

INRO 2008. Emme/2 and 3 software. Retrieved September 18, 2008, from *www.inro.ca/en/products/emme/index.php.*

Integral GIS, Inc. 2006. 4D GIS construction management. Retrieved August 25, 2006, from *www.integralgis.com/safeco.aspx.*

Intermodal Surface Transportation Efficiency Act (ISTEA) of 1991. (Public Law 102-240, December 18, 1991), 105 United States Statutes at Large, 1914-2207.

International Association of Public Participation. 2005. Public participation spectrum. Retrieved August 10, 2005, from *www.iap2.org/associations/4748/files/spectrum.pdf.*

Jankowski, P. 1995. Integrating geographic information systems and multiple criteria decision making methods. *International Journal of Geographical Information Systems, 9*(3), 251–273.

Jankowski, P., Lotov, A., and Gusev, D. 1999. Application of multicriteria trade-off approach to spatial decision making. In J.-C. Thill (Ed.), *GIS and Multiple Criteria Decision Making: A Geographic Information Science Perspective*. London: Ashgate, pp. 127–148.

Jankowski, P., and Nyerges, T. 2001a. *Geographic Information Systems for Group Decision Making*. London: Taylor and Francis.

Jankowski, P., and Nyerges, T. 2001b. GeoChoicePerspectives: A collaborative spatial decision support system. In D. L. Schmoldt, J. Kangas, G. Mendoza, and M. Pesonen (Eds.), *The Analytic Hierarchy Process in Natural Resource and Environmental Decision Making*. Dordrecht, the Netherlands: Kluwer Academic, pp. 253–268.

Jankowski, P., Robischon, S., Tuthill, D., Nyerges, T., and Ramsey, K. 2006. Design considerations and evaluation of a collaborative, spatio-temporal decision support system. *Transactions in GIS, 10*(3), 335–354.

Kates, R. K., Clark, W. C., Corell, R., Hall, J. M., Jaeger, C. C., Lowe, I., et al. 2001. Sustainability science. *Science, 292*, 641–642.

Kelly, E., and Becker, B. 2000. *Community Planning: An Introduction to the Comprehensive Plan*. Washington, DC: Island Press.

Kenney, R. 1992 *Value-focused thinking: A path to creative decision making*. Cambridge, MA: Harvard University Press.

Kenney, R., von Winterfeldt, D., and Eppel, T. 1990. Eliciting public values for complex policy decisions. *Management Science, 36*(9), 1011–1030.

Kent, W. 1984. A realistic look at data. *Database Engineering, 7*(4), 22–27.

King County. 2001. Siting the Brightwater treatment facilities. Retrieved December 7, 2006, from *dnr.metrokc.gov/wtd/brightwater/library/docs/siting-mar00.pdf*.

King County. 2002a. Concurrency Management Program, King County Department of Transportation. Retrieved August 15, 2002, from *www.metrokc.gov/kcdot/tp/concurr/conindex.htm*.

King County. 2002b. Concurrency Management Program, King County, Washington. *www.metrokc.gov/kcdot/tp/concurr/attachmentA32002.pdf*.

King County. 2004. *Siting the Brightwater Treatment Facilities, Phase 2 Summary* (page 22 Table 4). Seattle, WA: King County Department of Natural Resources.

Lai, S.-K. 1998. From organized anarchy to controlled structure: Effects of planning on the garbage can decision processes. *Environment and Planning B, 25*, 85–102.

Lai, S.-K. and Hopkins, L. D. 1989. The meanings of trade-offs in multi-attribute utility methods: A comparison. *Environment and Planning B: Planning and Design 16*, 155–170.

Lee, I. and Hopkins, L. D. 1995. Procedural expertise for efficient multiattribute evaluation: A procedural support strategy for CEA. *Journal of Planning, Education and Research, 14*(4), 225–268.

Lober, D. J. 1995, Autumn. Resolving the siting impasse: Modeling social and environmental locational criteria with a geographic information system. *American Planning Association Journal*, pp. 482–495.

Longley, P., Goodchild, M., Maguire, D., and Rhind, D. 2005. *Geographic Information Systems and Science*. Chichester, UK: Wiley.

Louisiana Department of Transportation and Development. 2006. Transportation Project Viewer. Retrieved August 24, 2006, from *dotdgis2.dotd.louisiana.gov/website/Construction/viewer.htm*.

Lowry, J., Miller, H., and Hepner, G. 1995. A GIS-based sensitivity analysis of community vulnerability to hazardous contaminants on the Mexico/US border. *Photogrammetric Engineering and Remote Sensing, 61*(11), 1347–1359.

Malczewski, J. 1999. *GIS and Multicriteria Decision Analysis.* New York: Wiley.

Martin, J. 1976. *Principles of Data-Base Management.* Englewood Cliffs, NJ: Prentice-Hall.

Martinez, C., and Wright, S. 2004. Design–build transportation and watershed planning meet in GIS. ESRI 2006 User Conference. Retrieved August 24, 2006, from *gis.esri.com/library/userconf/proc04/docs/pap2016.pdf.*

Maxwell, T. 2006. Spatial modeling environment. Retrieved May 8, 2006, from *www.uvm.edu/giee/sme3.*

Meyer, M., and Miller, E. 2001. *Urban Transportation Planning: A Decision-Oriented Approach.* London: McGraw-Hill.

Miles, S. 2004. *Comprehensive Areal Model for Earthquake-Induced Landslides.* Unpublished doctoral dissertation, Department of Geography, University of Washington, Seattle.

Mitchell B. (Ed.). 1990. *Integrated water management,* London: Belhaven Press.

Munn R. E., 1975, *Environmental Impact Assessment Principles and Procedures: SCOPE 5.* Toronto: International Council of Scientific Unions, Scientific Committee on Problems of the Environment.

National Environmental Policy Act (NEPA). 1970. Retrieved August 7, 2006, from *www.epa.gov/compliance/nepa.*

National Institute for Standards and Technology. 1994. Federal Information Processing Standard 173-1, Spatial Data Transfer Standard. Gaithersburg, MD: Author.

National Oceanic and Atmospheric Administration, Coastal Services Center (NOAA CSC). 2006a. Alternatives for coastal development: One site, three scenarios. Retrieved August 22, 2006, from *www.csc.noaa.gov/alternatives/overview.html.*

National Oceanic and Atmospheric Administration, Coastal Services Center (NOAA CSC). 2006b. Audience analysis. Retrieved August 22, 2006, from *www.csc.noaa.gov/alternatives/audience_analysis.pdf.*

National Oceanic and Atmospheric Administration, Coastal Services Center (NOAA CSC). 2006c. Indicators. Retrieved August 22, 2006, from *www.csc.noaa.gov/alternatives/indicators.html.*

National Oceanic and Atmospheric Administration, Coastal Services Center (NOAA CSC). 2006d. Indicator methods. Retrieved August 22, 2006, from *www.csc.noaa.gov/alternatives/indicatormethods.html.*

National Oceanic and Atmospheric Administration, Coastal Services Center (NOAA CSC). 2006e. Scenarios. Retrieved August 22, 2006, from *www.csc.noaa.gov/alternatives/scenarios.html.*

National Research Council. 1996. *Understanding Risk: Informing Decisions in a Democratic Society.* Washington, DC: National Academy Press.

National Research Council. 1999. *Our Common Journey: A Transition Toward Sustainability.* Washington, DC: National Academies Press.

National Research Council. 2006. *Learning to Think Spatially: GIS as a Support System in the K–12 Curriculum.* Washington, DC: National Academies Press.

National Resources Conservation Service. 2006a. Soils data. Retrieved July 31, 2006, from *soildatamart.nrcs.usda.gov.*

National Resource Conservation Service. 2006b. Soils maps. Retrieved July 31, 2006, from *websoilsurvey.nrcs.usda.gov/app.*

New Jersey Department of Environmental Protection. 2006. Blueprint for intelligent growth: Land use permitting policy. Retrieved May 3, 2006, from *www.state.nj.us/dep/commissioner/policy/pdir2003-001.htm*.

New Jersey Department of Transportation. 2006. Statewide Transportation Improvement Program. Retrieved June 17, 2006, from *www.state.nj.us/transportation/capital/stip0608*.

Niemann, B. J. 1989. Improved analytical functionality: Modernizing land administration, planning, management, and policy analysis. In P. Brown and D. Moyer (Eds.), *Multipurpose Land Information System Guidebook*. Washington, DC: NOAA, National Geodetic Survey, Chapter 11.

Nyerges, T. 1989a. Schema integration analysis for GIS database development. *International Journal of Geographic Information Science*, 3(2), 153–183.

Nyerges, T. 1989b. Information integration for multipurpose land information systems. *Urban and Regional Information Systems Association Journal*, 1(1), 28–39.

Nyerges, T. 1990, March. Locational referencing and highway segmentation in a geographic information system. *ITE Journal*, pp. 27–31.

Nyerges, T. 1991a. Analytical map use. *Cartography and Geographic Information Systems*, 18(1), 11–22.

Nyerges, T. 1991b. Geographic information abstractions: Conceptual clarity for geographic modeling. *Environment and Planning A*, 23, 1483–1499.

Nyerges, T. 1993. Understanding the scope of GIS: Its relationship to environmental modeling. In M. Goodchild, L. Steyaert, and B. Parks (Eds.), *Environmental Modeling with Geographic Information Systems*. London: Oxford University Press, pp. 75–93.

Nyerges, T. 1995. GIS support for urban and regional transportation analysis. In Susan Hanson (Ed.), *The Geography of Urban Transportation* (2nd ed.). New York: Guilford Press, pp. 240–265.

Nyerges, T. 2005. Scaling-up as a grand challenge for public participation GIS. Retrieved September 20, 2005, from *www.directionsmag.com/article.php?article_id=1965*.

Nyerges, T., Barndt, M., and Brooks, K. 1997. Public participation geographic information systems. *AUTOCARTO 13* Conference Proceedings (Seattle, WA). Bethesda, MD: American Congress on Surveying and Mapping, pp. 224–233.

Nyerges, T., and Jankowski, P. 1997. Enhanced adaptive structuration theory: A theory of GIS-supported collaborative decision making. *Geographical Systems*, 4(3), 225–259.

Nyerges, T., Montejano, R., Oshiro, C., and Dadswell, M. 1998. Group-based geographic information systems for transportation site selection. *Transportation Research C: Emerging Technologies*, 5(6), 349–369.

Oloufa, A., Eltahan, A., and Papacostas, C. 1994. Integrated GIS for construction site investigation. *Journal of Construction Engineering and Management*, 120(1), 211–222.

Ortolano, L., and Sheperd, A. 1995. Environmental impact assessment: Challenges and opportunities. *Impact Assessment*, 3(1), 3–30.

Parenteau, R. 1988. *Public Participation in Environmental Decision Making*. Ottawa: Federal Environmental Assessment Review Office.

Parker, S., Parker, D., and Stader, T. 1995. A geographic information system to predict soil erosion potential in rural transportation construction project. Retrieved August 24, 2006, from *ntl.bts.gov/lib/10000/10600/10661/mbtc1005.pdf*.

Pawlak, Z., and Slowinski, R. 1994. Rough set approach to multi-attribute decision analysis. *European Journal of Operational Research*, 72, 443–459.

Peng, Z.-R., and Tsou, M.-H. 2003. *Internet GIS*. Hoboken, NJ: Wiley.

Pew Commission. 2002. Oceans Commission Science Report (32 pp.). Retrieved August 22, 2006, from *www.nga.org/cda/files/102203wavescolgan.pdf*.

Poole, M. S., Seibold, D. R., and McPhee, R. D. 1985. Group decision-making as a structurational process. *Quarterly Journal of Speech*, 71, 74–102.

Pope, C. 1999. Solving sprawl. Retrieved April 8, 2006, from *www.sierraclub.org/sprawl/report99*.

Porter, D. R. 1997. *Managing Growth in America's Communities*. Washington, DC: Island Press.

Puget Sound Regional Council. 2001. *Destination 2030*, CD-ROM, May 24, 2001, Seattle, Washington: Puget Sound Regional Council.

Puget Sound Regional Council. 2006a. VISION 2020 Update. Retrieved May 3, 2006, from *www.psrc.org/projects/vision/index.htm*.

Puget Sound Regional Council. 2006b. Destination 2030. Retrieved May 3, 2006, from *www.psrc.org/projects/mtp/index.htm*.

Puget Sound Regional Council. 2006c. Freight action strategy. Retrieved August 31, 2006, from *www.psrc.org/fastcorridor/fastbrochure.pdf*.

Purdue University. 2006. Revised Universal Soil Loss Equation 2 software. Retrieved August 25, 2006, from *fargo.nserl.purdue.edu/rusle2_dataweb/RUSLE2_Index.htm*.

Ramsey, K. 2004. *Stakeholder Involvement and Complex Decision Making: A Case Study into the Design and Implementation of a GIS for Supporting Local Water Resource Management*. Unpublished master's thesis, Department of Geography, University of Washington, Seattle.

Randolph, J. 2004. *Environmental Land Use Planning and Management*. Washington, DC: Island Press.

Rees, W. 1998. Understanding sustainable development. In B. Hamm and P. K. Muttagi (Eds.), *Sustainable Development and the Future of Cities*. London: Intermediate Technology Publications, pp. 19–42.

Rumbaugh, J., Jacobson, I., and Booch, G. 1999. *The Unified Modeling Language Reference Manual*. Reading, MA: Addison-Wesley.

Saaty, T. L. 1980. *The Analytic Hierarchy Process*. New York: McGraw-Hill.

Sadler, B. 1996. *Environmental Assessment in a Changing World: Evaluating Practice to Improve Performance* (International Study of the Effectiveness of Environmental Assessment). Ottawa: Canadian Environmental Assessment Agency and International Association for Impact Assessment.

Safe, Accountable, Flexible, Efficient Transportation Equity Act: A Legacy for Users (SAFTEA-LU, Public Law 109-203). 2005. California Transportation Commission. Retrieved June 16, 2006, from *www.dot.ca.gov/hq/transprog/index.htm*.

Saisana, M., Saltelli, A., and Tarantola, S. 2005. Uncertainty and sensitivity analysis techniques as tools for the quality assessment of composite indicators. *Journal of Royal Statistical Society A*, *168*(2), 307–323.

Saltelli, A., Chan, K., and Scott, E. M. (Eds.). 2001. *Sensitivity Analysis*. New York: Wiley.

Sayer, A. 1984. *Method in Social Science*. London: Hutchinson.

Shiffer, M. J. 1995. Interactive multimedia planning support: Moving from stand-alone systems to the World Wide Web. *Environment and Planning B*, *22*, 649–664.

Shiffer, M. J. 1998. Multimedia GIS for planning support and public discourse. *Cartography and GIS*, *25*(2), 89–94.

Shiffer, M. J. 2002. Spatial multimedia representations to support community participation. In

W. Craig, T. Harris, and D. Weiner (Eds.), *Community Participation and Geographic Information Systems*. London: Taylor and Francis, pp. 309–319.

Simon, H. 1977. *The New Science of Management Decisions* (3rd ed.). Englewood Cliffs, NJ: Prentice-Hall.

Smith, L. G. 1982. Alternative mechanisms for public participation in environmental policy-making. *Environments 14*(3), 21–34.

Steinitz, C. 1990. A framework for theory applicable to the education of landscape architects (and other design professionals). *Landscape Journal, 9*(2), 136–143.

Steinitz, C. 1996. Biodiversity and landscape planning : Alternative futures for the region of Camp Pendleton, California, Strategic Environmental Research and Development Program (U.S.), U. S. Army. Retrieved September 2, 2006, from *www.gsd.harvard.edu/studios/brc/brc.html*

Steinitz, C., Arias, H., Bassett, S., Flaxman, M., Goode, T., Maddock III, T., et al. 2003. *Alternative futures for changing landscapes: The upper San Pedro River basin in Arizona and Sonora.* Washington, DC: Island Press.

Steinitz, C., Faris, R., Flaxman, M., Vargas-Moreno, J. C., Huang, G., Lu, S.-Y., et al. 2005. Alternative futures for the region of La Paz, Baja California Sur, Mexico. Retrieved March 1, 2009, from *Projects.Gsd.Harvard.Edu/Lapaz/Index.html*.

Talen, E. 1998. Visualizing fairness: Equity maps for planners. *Journal of the American Planning Association, 64*(1), 64–75.

Task Committee on Sustainability Criteria. 1998. *Sustainability Criteria for Water Resource Systems* (Water Resources Planning and Management Division, American Society of Civil Engineers and Working Group UNESCO/IHP IV Project M-4.3). Reston, VA: American Society of Civil Engineers Publications.

Texas Water Development Board. 2004. Water resources planning and information. Retrieved December 10, 2004, from *www.twdb.state.tx.us/rwpg/what-is-rwp.asp#top*.

Tobler, W. 1979. Transformational View of Cartography. *American Cartographer, 6*(2), 101–106. Retrieved November 1, 2006, from *www.geog.ucsb.edu/~tobler/publications/pdf_docs/cartography/maptrnsfrm.pdf*.

Town of Newstead. 2006, Spring. For Newstead, New York, GIS improves construction project management. *ESRI News*. Retrieved August 25, 2006, from *www.esri.com/news/arcnews/spring06articles/for-newstead-ny.html*.

Transportation Equity Act for the 21st Century (TEA-21). 1998. Public Law 105-178, Title 23 Part 450.324, subpart C.

Transportation Research Board. 1985. *Highway Capacity Manual*. Washington, DC: National Academies Press.

Tuthill, D. 2002. *Utilization of Emerging Geo-Spatial Technologies in the Implementation of Conjunctive Administration of Surface and Ground Water.* Unpublished Ph.D. dissertation, Department of Civil Engineering, University of Idaho, Boise.

Ullman, J. 1980. *Principles of Database Systems.* New York: Freeman.

U.S. Advisory Commission on Intergovernmental Relations. 1997. *Planning Progress: Addressing ISTEA Reauthorization in Metropolitan Planning Areas.* Washington, DC: U.S. Department of Transportation.

U.S. Army Corps of Engineers. 2009. Lower Snake River juvenile salmon migration feasibility Study Ch 7. Social Impact Analysis. Retrieved March 31, 2009, from *www.nww.usace.army.mil/lsr/REPORTS/social/impacts.pdf*.

U.S. Bureau of the Census. 2003. State and County QuickFacts. *United States: Population, Percent Change, 1990 to 2000.* Retrieved August 20, 2008, from *quickfacts.census.gov/qfd/states/00000. html.*

U.S. Environmental Protection Agency. 2006. Smart Growth Index. Retrieved August 22, 2006, from *www.epa.gov/smartgrowth/topics/sg_index.htm.*

U.S. Environmental Protection Agency (US EPA). 2009. Introduction to the Watershed Planning Process, from EPA's online watershed academy. Retrieved March 31, 2009, from *www.epa. gov/watertrain/planning/index.htm*

U.S. Geological Survey. 2006a. National Hydrography Dataset. Retrieved July 31, 2006, from *nhd. usgs.gov.*

U.S. Geological Survey. 2006b. National Hydrography Dataset (ArcView Toolkit). Retrieved July 31, 2006, from *nhd.usgs.gov/tools.html#nhdgeoshp.*

U.S. Geological Survey. 2009. *National handbook of recommended methods for water data acquisition.* Retrieved June 1, 2009, from *pubs.usgs.gov/chapter11.*

Vanclay, F., and Bronstein, D. 1995. *Environment and Social Impact Assessments.* London: Wiley.

Voogd, H. 1983. *Multicriteria Evaluation for Urban and Regional Planning.* London: Pion.

Wachs, M., and Schofer, J. L. 1969. Abstract values and concrete highways. *Traffic Quarterly,* 23(1), 133–145.

Wachs, M., and Schofer, J. L. 1976. Structuring a participatory decision framework for public systems. In *Proceedings: Specialty Conference in Human Factors in Civil Engineering Planning, Design, And Education.* Buffalo: Civil Engineering Department, State University of New York, pp. 225–256.

Washington State. 1990. Growth Management Act (updated 1991). Retrieved March 1, 2007, from *historylink.org/essays/output.cfm?file_id=7759.*

Washington State. 2006. Citizens building a better Washington. Retrieved April 8, 2006, from *cted.wa.gov/_cted/documents/id_1014_publications.pdf.*

Washington State. 2009. Growth management. Retrieved April 2, 2009, from *www.cted.wa.gov/_ CTED/documents/ID_892_Publications.pdf*

Washington State Department of Ecology. 2006. Water resource inventory area maps. Retrieved August 25, 2006, from *www.ecy.wa.gov/services/gis/maps/wria/wria.htm.*

Washington State Department of Ecology. 2007. 2007–2009 Strategic Plan. Retrieved March 28, 2008, from *www.ecy.wa.gov/pubs/0601005.pdf.*

Washington State Department of Transportation. 2006. Washington transportation framework for GIS. Retrieved November 8, 2006, from *www.wsdot.wa.gov/mapsdata/transframework/ default.htm.*

Washington State Environmental Protection Act. 1970. Retrieved August 7, 2006, from *www.ecy. wa.gov/programs/sea/sepa/e-review.html.*

Washington State Housing Finance Commission. 2006a. Retrieved September 1, 2006, from *www.wshfc.org.*

Washington State Housing Finance Commission. 2006b. Retrieved September 1, 2006, from *www.wshfc.org/tax-credits.*

Washington State Housing Finance Commission. 2006c. Retrieved September 1, 2006, from *www.wshfc.org/tax-credits/2005application/index.htm.*

Washington State Housing Finance Commission. 2006d. Retrieved September 1, 2006, from *www.wshfc.org/tax-credits/2005application/j-qct-dda.pdf.*

Washington State Housing Finance Commission. 2006e. Retrieved September 1, 2006, from *www.wshfc.org/tax-credits/2005application/b-qap.pdf.*

White, R. 2002. *Building the Ecological City.* Boca Raton, FL: CRC Press.

World Commission on Environment and Development. 1987. Published as *Our Common Future,* Annex to General Assembly document 42/1827, *Development and International Co-operation: Environment* August 2, 1987. Retrieved March 28, 2008, from *www.un-documents.net/a42r187. htm.*

Wyoming State Water Plan. 2004. Summary of the state water planning process. Retrieved December 10, 2004, from *waterplan.state.wy.us.*

Young, R. K., Zhong, T., Lowry, M., Rutherford, G. S., and Nyerges, T. L. 2007. An analytic–deliberative online framework for transportation programming. *International Journal of Technology, Knowledge and Society,* 3(2), 89–98.

Younger, K., and O'Neill, C. 1998. Making the connection: The Transportation Improvement Program and the long-range plan (Transportation Research Record No. 1617). In *Land Use and Transportation Planning and Programming Applications.* Washington, DC: National Academies of Sciences, pp. 118–121.

Yuan, M., and Stewart Hornsby, K. 2007. *Computation and Visualization for the Understanding of Dynamics in Geographic Domains: A Research Agenda.* Boca Raton, FL: CRC Press.

Zeiler, M. 1999. *Modeling Our World.* Redlands, CA: ESRI Press.

Zhong, D., Li, J., Zhu, H., and Song, L. 2004. Geographic information system-based visual simulation methodology and its application in concrete dam construction processes. *Journal of Construction, Engineering and Management,* 130(5), 742–750.

Index

About the Authors

Timothy L. Nyerges is Professor of Geography and Affiliate Professor of the Water Center at the University of Washington. He teaches introductory, intermediate, and advanced courses to undergraduate and graduate students in GIS using ArcGIS software and systems design using open-source GIS software. His research focuses on public participation decision support using both workstation and online platforms for urban–regional land, transportation, and water resources activity.

Piotr Jankowski is Professor of Geography and Co-Director of the Center for Earth Systems Analysis Research at San Diego State University. He teaches undergraduate and graduate courses in GIS and spatial decision analysis. His research focuses on spatial decision support systems, participatory GIS, spatial optimization, and modeling methods.